Dynamics of the Coastal Zone

Special Issue Editors

Matteo Postacchini
Alessandro Romano

MDPI • Basel • Beijing • Wuhan • Barcelona • Belgrade • Manchester • Tokyo • Cluj • Tianjin

Special Issue Editors

Matteo Postacchini
Università Politecnica delle Marche
Italy

Alessandro Romano
"Sapienza" University of Rome
Italy

Editorial Office
MDPI
St. Alban-Anlage 66
4052 Basel, Switzerland

This is a reprint of articles from the Special Issue published online in the open access journal *Journal of Marine Science and Engineering* (ISSN 2077-1312) (available at: https://www.mdpi.com/journal/jmse/special_issues/bz_dynamics_coastal_zone).

For citation purposes, cite each article independently as indicated on the article page online and as indicated below:

LastName, A.A.; LastName, B.B.; LastName, C.C. Article Title. *Journal Name* **Year**, *Article Number, Page Range*.

ISBN 978-3-03928-484-9 (Pbk)
ISBN 978-3-03928-485-6 (PDF)

Cover image courtesy of Alessandro Romano.

Dynamics of the Coastal Zone

Contents

About the Special Issue Editors

Matteo Postacchini (Ph.D.): Assistant Professor since 2016 at the Department of Civil and Building Engineering, and Architecture of the Università Politecnica delle Marche (Ancona, Italy). His research experience is related to numerical modeling (programming and implementation of solvers for the description of nearshore/shallow-water flows); field experience in coastal, riverine, and estuarine environments; and physical modeling of traditional and novel structures for coastal protection through laboratory tests. He has participated in several international projects and worked in several universities as a visiting researcher: Scripps Institution of Oceanography (California, USA), EPFL (Switzerland), Bordeaux 1 (France), Universitat Politècnica de Catalunya (Spain), among others. He is author/co-author of several ISI/Scopus journals. He acts as a reviewer for more than 25 journals and is an Editorial Board Member of the Journal of Marine Science and Engineering (MDPI).

Alessandro Romano (Ph.D.): Postdoc researcher since 2013 at Roma Tre University (Rome, Italy) and later at Sapienza University of Rome (Rome, Italy). His research activity is mainly related to physical and numerical modelling of landslide-generated tsunamis, wave–structure interaction (wave overtopping and impact), numerical modelling of coastal hydrodynamics, management of coastal defence, and generation techniques of waves in laboratory. He hasz participated in several national and international research projects (e.g. EU project MERMAID) and has been a visiting researcher at renowned research institutes: Deltares (Delft, The Netherlands) and IHCantabria (Santander, Spain). He has been the National Secretary of PIANC Italy and member of the Commission Rapporteur at the Italian High Council for Public Works. He is author of more than 40 scientific papers (32 indexed in Scopus) in peer-reviewed international journals, has attended many national and international conferences, and is a member of the Scientific Committee of the International Conference Coastal Structures 2019 (ASCE) and the International Conference Meddays 2018 (PIANC).

Journal of
Marine Science and Engineering

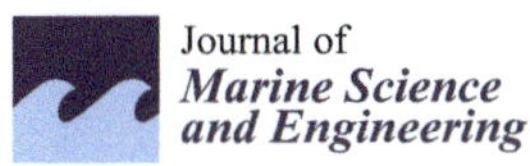

Editorial

Dynamics of the Coastal Zone

Matteo Postacchini [1],* and Alessandro Romano [2]

[1] Department of Civil and Building Engineering and Architecture, Università Politecnica delle Marche, via Brecce Bianche 12, I-60131 Ancona, Italy
[2] Department of Civil, Constructional and Environmental Engineering (DICEA), Sapienza University of Rome, via Eudossiana 18, 00184 Rome, Italy; alessandro.romano@uniroma1.it
* Correspondence: m.postacchini@staff.univpm.it

Received: 5 December 2019; Accepted: 5 December 2019; Published: 9 December 2019

Keywords: coastal region; surf zone; swash zone; beach erosion; hydrodynamics; morphodynamics; laboratory experiments; analytical and numerical modeling; statistical methods; climate changes

---

1. Overview

The coastal zone hosts many human activities and interests, which have significantly increased in the last few decades. However, climate change may have destabilizing effects on such activities all over the world: sea level rise and the increase in the magnitude and frequency of storm events can severely affect beaches and coastal structures, with negative consequences and dramatic impacts for coastal communities from different (e.g., ecological, recreational, environmental) points of view. These aspects add to typical coastal problems, such as flooding and beach erosion, among others, already leading to large economic losses and human fatalities. Analytical, numerical, and physical modeling are thus fundamental approaches to be jointly used by scientists for an exhaustive understanding of the nearshore region in the present and future environment. For this purpose, innovative tools and technologies may help to provide a more detailed interpretation of the coastal processes, in terms of hydrodynamics, sediment transport, bed morphology, and their interaction with coastal structures.

The present Special Issue (SI), titled "Dynamics of the Coastal Zone" and hosted by the *Journal of Marine Science and Engineering*, aims at collecting the most recent contributions focusing on the nearshore region. These deal with different modeling approaches and different analyzed processes, while spanning among several time and spatial scales. Specifically, some of the presented studies analyze the main results coming from notable laboratory facilities [1–4], while the other contributions pertain to the use of advanced modeling approaches [5–9]. Due to the complexity of the coastal environment, some of the SI manuscripts are mainly related to the hydrodynamic processes [1,3,4,6,7,9], while others to the analysis of sediment-transport-related issues [2,5,8]. A further classification concerns the scales: while some works investigate relatively short processes, related to either the wave/intra-wave analysis [1,3,7] or the storm/event scale [2,4,6,9], the remaining works investigate long term effects in the nearshore region [5,8]. The details of the individual contributions are summarized in the following section.

2. Contributions

De Serio and Mossa [1] describe important experimental findings related to the wave-breaking process. They used a fixed sloping beach and different wave conditions. Both spilling and plunging breakers were analyzed in terms of water elevation and velocity distribution shoreward of the breaking location. Observations have allowed understanding turbulent features and coherent motions generated during the analyzed processes.

Eichentopf et al. [2] illustrate laboratory tests aimed at understanding the morphological response of a sandy beach subject to the same wave condition, when different initial profiles are used. All configurations, although starting from different initial conditions, reach the same equilibrium state, this demonstrating that the same wave forcing generates different sediment transport patterns, with the largest beach changes and hydrodynamic differences during the test start. Further, large breaker bars promote energy dissipation and limit shore erosion, while large berms, combined with small or no bars, promote shore retreat.

The laboratory experiments discussed in Howe et al. [3] are focused on the swash zone bed shear stress, measured using both medium and prototype scale configurations. It was observed that peak shear stress coincides with the arrival of uprush swash fronts, while the friction factor slightly changes during the wave cycle and decreases with Reynolds number on smooth slopes. Such estimated friction factors are larger than expected, when plotted on Moody or wave friction diagrams.

Williams et al. [4] focus on the assessment of the uncertainties of wave overtopping occurring at different types of coastal structures. Their laboratory tests are aimed at describing the variation in the main overtopping measures when different wave time series, generated from the same spectrum, are used. Many wave conditions were tested using two different structures, one characterized by a smooth slope, the other by a vertical wall. Large variations of the overtopping discharge are observed when different time series are used.

The work by Albernaz et al. [5] deals with the current and wave induced sediment transport, which requires a suitable parameterization of the wave orbital velocities. This is particularly important when long period simulations are performed; hence, a new parameterization is presented, which accounts for both skewness and asymmetry. Such parameterization was thus implemented in an existing numerical model (Delft3D) and provided suitable coastline predictions in long time periods.

Gallerano et al. [6] describe a three-dimensional solver based on the integral contravariant formulation of the Navier–Stokes equations. The authors analyze the hydrodynamics induced by a submerged breakwater subjected to the wave field and located in a coastal area characterized by a curvilinear shoreline. The aim is that of understanding the circulation patterns, as well as the change in the hydrodynamic conditions promoted by the structure.

In their paper, Moura and Baldock [7] introduce a simple numerical model aimed at evaluating the wave shape during shoaling with concurrent radiation of free long waves. Comparisons with simulations from an existing solver are illustrated, also to investigate the dissipation of free and forced long waves in the surf zone. Results from both models used suggest that both the growth rate and lag of the long wave behind the forcing are frequency dependent, in agreement with the literature and more complex evolution models.

The work by O'Grady et al. [8] provides an analysis of future changes at Lakes Entrance (Australia), through data downscaling from global and regional climate models to force a local climate model at the beach scale. The future sediment transport induced by such modeling has been obtained using empirical and detailed models. The authors introduce a new downscaling method and observe that modeled changes to wave transport are an order of magnitude larger than changes from storm-tide current and mean sea level changes.

Postacchini and Ludeno [9] combine numerical simulations with a typical approach to estimate sea data from X-band radar signals. Such a combination was applied for coastal inundation purposes. First, the application of the radar approach to simulated data allowed estimating both the wave field and bathymetry. Then, such results were used to run numerical simulations of coastal inundation. Results coming from the two steps above suggest that the proposed combination may be successfully used for several coastal purposes (e.g., hazard mapping, warning systems).

The above summaries briefly recap the SI contents, which are related to the main processes occurring in the nearshore region, here analyzed using various techniques and spanning through different scales.

Dr. Matteo Postacchini and Dr. Alessandro Romano: Guest Editors of "Dynamics of the Coastal Zone".

Funding: This research received no external funding.

Acknowledgments: All contributing authors and reviewers are thanked for their efforts.

Conflicts of Interest: The authors declare no conflict of interest.

References

1. De Serio, F.; Mossa, M. Experimental Observations of Turbulent Events in the Surfzone. *J. Mar. Sci. Eng.* **2019**, *7*. [CrossRef]
2. Eichentopf, S.; van der Zanden, J.; Cáceres, I.; Alsina, J.M. Beach Profile Evolution towards Equilibrium from Varying Initial Morphologies. *J. Mar. Sci. Eng.* **2019**, *7*. [CrossRef]
3. Howe, D.; Blenkinsopp, C.E.; Turner, I.L.; Baldock, T.E.; Puleo, J.A. Direct Measurements of Bed Shear Stress under Swash Flows on Steep Laboratory Slopes at Medium to Prototype Scales. *J. Mar. Sci. Eng.* **2019**, *7*. [CrossRef]
4. Williams, H.E.; Briganti, R.; Romano, A.; Dodd, N. Experimental Analysis of Wave Overtopping: A New Small Scale Laboratory Dataset for the Assessment of Uncertainty for Smooth Sloped and Vertical Coastal Structures. *J. Mar. Sci. Eng.* **2019**, *7*. [CrossRef]
5. Albernaz, M.B.; Ruessink, G.; Jagers, H.R.A.B.; Kleinhans, M.G. Effects of Wave Orbital Velocity Parameterization on Nearshore Sediment Transport and Decadal Morphodynamics. *J. Mar. Sci. Eng.* **2019**, *7*. [CrossRef]
6. Gallerano, F.; Cannata, G.; Palleschi, F. Hydrodynamic Effects Produced by Submerged Breakwaters in a Coastal Area with a Curvilinear Shoreline. *J. Mar. Sci. Eng.* **2019**, *7*. [CrossRef]
7. Moura, T.; Baldock, T.E. The Influence of Free Long Wave Generation on the Shoaling of Forced Infragravity Waves. *J. Mar. Sci. Eng.* **2019**, *7*. [CrossRef]
8. O'Grady, J.; Babanin, A.; McInnes, K. Downscaling Future Longshore Sediment Transport in South Eastern Australia. *J. Mar. Sci. Eng.* **2019**, *7*. [CrossRef]
9. Postacchini, M.; Ludeno, G. Combining Numerical Simulations and Normalized Scalar Product Strategy: A New Tool for Predicting Beach Inundation. *J. Mar. Sci. Eng.* **2019**, *7*. [CrossRef]

Article

Beach Profile Evolution towards Equilibrium from Varying Initial Morphologies

Sonja Eichentopf [1,*], Joep van der Zanden [2,3] , Iván Cáceres [4] and José M. Alsina [4]

[1]   Fluid Mechanics Section, Department of Civil and Environmental Engineering, Imperial College London, London SW7 2AZ, UK
[2]   Marine and Fluvial Systems Group, University of Twente, Drienerlolaan 5, 7522 NB Enschede, The Netherlands; j.v.d.zanden@marin.nl
[3]   Offshore Department, Maritime Research Institute Netherlands, Haagsteeg 2, 6708 PM Wageningen, The Netherlands
[4]   Laboratori d'Enginyeria Marítima, Universitat Politècnica de Catalunya, 08034 Barcelona, Spain; i.caceres@upc.edu (I.C.); jose.alsina@upc.edu (J.M.A.)
*   Correspondence: sonja.eichentopf16@imperial.ac.uk

Received: 16 October 2019; Accepted: 5 November 2019; Published: 9 November 2019

Abstract: The evolution of different initial beach profiles towards the same final beach configuration is investigated based on large-scale experimental data. The same wave condition was performed three times, each time starting from a different initial profile morphology. The three different initial profiles are an intermediate energy profile with an offshore bar and a small swash berm, a plane profile and a low energy profile with a large berm. The three cases evolve towards the same final (equilibrium) profile determined by the same wave condition. This implies that the same wave condition generates different sediment transport patterns. Largest beach changes and differences in hydrodynamics occur in the beginning of the experimental cases, highlighting the coupling between morphology and hydrodynamics for beach evolution towards the same profile. The coupling between morphology and hydrodynamics that leads to the same final beach profile is associated with differences in sediment transport in the surf and swash zone, and is explained by the presence of bar and berm features. A large breaker bar and concave profile promote wave energy dissipation and reduce the magnitudes of the mean near-bed flow velocity close to the shoreline limiting shoreline erosion. In contrast, a beach profile with reflective features, such as a large berm and a small or no bar, increases negative velocity magnitudes at the berm toe promoting shoreline retreat. The findings are summarised in a conceptual model that describes how the beach changes towards equilibrium from two different initial morphologies.

Keywords: beach equilibrium; initial morphology; large-scale experiments; beach erosion; beach recovery; sediment transport

---

1. Introduction

It is a widely accepted concept that beaches (hereinafter defined as the entire active movable bed from the shoaling to the swash zone [1]) evolve towards an equilibrium configuration under constant wave action for a sufficient duration. It states that, for a given wave condition, the beach morphology and the hydrodynamics develop together towards a stable, i.e., equilibrium, condition with no sediment transport gradients [2,3]. In fact, in most cases, a strict equilibrium condition is not reached and hence equilibrium conditions often refer to a quasi-equilibrium where beach changes are not exactly zero but close to zero [3]. Formulations of equilibrium beach profiles have been established and they have been fundamental for the development of beach evolution models [4–8].

Wright and Short [9] presented the concept of the equilibrium beach states that result from the dominant surf zone wave forcing. Different beach states can be distinguished by morphological beach features, of which the bar and the berm are two of the most striking in the two-dimensional plane. A bar and a small onshore berm are generally associated with higher energy beach states [9] where the bar promotes wave breaking and wave energy dissipation [10–14]. A large and wide berm is generally associated with low energy beaches [9]. The presence of a berm has been linked to berm overwash and sediment accumulation on the berm crest [15] and to horizontal sediment advection and the consequent cross-shore distribution of sediment transport [16–18].

In terms of the beach evolution towards equilibrium, important influencing factors are the wave conditions and their duration, as well as the initial beach configuration (e.g., [7,14,19–21]). Usually, the wave condition determines the equilibrium profile, towards which the beach evolves, while the duration of the wave forcing determines how close the beach profile is to the final equilibrium configuration (e.g., [4,7,19,21]). Consequently, depending on the initial morphology, a given wave condition can cause important differences in sediment transport which can lead to opposite signs in bulk sediment transport, i.e., the same wave condition can generate offshore or onshore sediment transport for different initial morphologies. This highlights that the terms 'erosion' and 'accretion' cannot be associated by default with high and low energy wave conditions, respectively, but that also the initial morphology needs to be considered [7,14,21].

In relation to the influence of the initial beach morphology, the availability of sediment along the profile is also important for an equilibrium profile to develop. Baldock et al. [14] and Birrien et al. [20] investigated morphological hysteresis where the same wave condition produced different equilibrium beach profiles, mainly because sediment becomes stranded offshore by storm conditions, making it unavailable for subsequent lower waves. Therefore, subsequent waves eroded sediment from around the shoreline generating a new equilibrium profile for these wave conditions. As a result, a certain wave energy can be required to mobilise stranded offshore sediments and to maintain the active beach profile [22,23].

Although the equilibrium concept is widely accepted in coastal morphodynamics, details of the coupling between hydrodynamics and morphology leading to an equilibrium profile are not fully understood. Eichentopf et al. [13] studied the coupling between the same initial morphology and different low energy wave conditions and identified important differences in the bar migration pattern. However, no data were available for the opposite case, where the initial morphology varies for the same wave condition. More detailed knowledge of the influence of the initial profile on equilibrium beach evolution is highly important to understand different beach changes generated by the same wave conditions which can ultimately be of interest for coastal management practices. This is, for instance, relevant in the context of beach nourishments which alter the beach profile for the same wave forcing [24]. It is also relevant in the recently growing research area of storm sequencing and beach response [25] as storms within sequences make landfall on different initial morphologies. Recent studies have highlighted beach equilibrium evolution under storm sequence forcing, but details of the resulting coupling between hydrodynamics and the varying morphologies are not well understood [21,26,27].

The aim of the present work is to investigate how different initial beach morphologies evolve towards the same final (equilibrium) profile under the same wave condition using a recently obtained large-scale laboratory data set of beach profile and hydrodynamic measurements. Specifically, it expands the work presented by Eichentopf et al. [21] by studying the observed equilibrium beach evolution of a specific wave condition starting from three different initial beach morphologies in detail.

This paper is organised as follows. Section 2 describes the experimental setup, including the wave condition, the measurements and the initial morphological conditions. The data analysis follows in Section 3. Results comprise the profile evolution, sediment transport and hydrodynamics and are presented in Section 4 followed by a conceptual model and discussion in Section 5. Section 6 concludes the paper.

2. Experimental Setup

The data investigated in the present work were acquired within the HYDRALAB$^+$ transnational access project RESIST. These experiments were performed in the large-scale wave flume (CIEM flume) at the Universitat Politècnica de Catalunya which is 100 m long, 3 m wide and 4.5 m deep. The flume is equipped with a wedge-type wave maker. A complete description of the experimental setup, the measuring programme and the wave conditions can be found in Eichentopf et al. [28]. Here, we describe the aspects that are most relevant for the present work.

In the RESIST experiments, two high energy (E1 and E2) and three low energy wave conditions (A1, A2 and A3) were combined into three sequences of alternating high-low energy wave conditions. At the beginning of each sequence, a benchmark case (condition B) with a significant wave height of 0.42 m and a peak period of 4 s was run for 30 min to homogenise and compact the manually shaped profile. Table 1 shows the order of the wave conditions with their duration and dimensionless sediment settling velocity Ω, which is calculated as:

$$\Omega = \frac{H_s}{w_s \cdot T_p} \tag{1}$$

where H_s (m) is the wave height, T_p (s) the wave period and w_s (m/s) the sediment settling velocity, which is 0.034 m/s. The reader is referred to Eichentopf et al. [21,28] for details of each wave condition.

Each of these sequences commenced from a 1/15 sloped, handmade initial beach profile. The shoreline of this plane profile, calculated as the intercept between the profile and the still water level (SWL), of each sequence is used as the origin of the x–z coordinate system. x corresponds to the horizontal direction and is negative towards the wave paddle; z presents the vertical coordinate and is positive upwards. The beach consisted of commercial sand with a narrow grain size distribution and a mean grain size $d_{50} = 0.25$ mm.

Table 1. Sequences of wave conditions, their duration and dimensionless sediment settling velocity Ω.

Sequence 1			Sequence 2			Sequence 3		
Condition	Duration (min)	Ω (-)	Condition	Duration (min)	Ω (-)	Condition	Duration (min)	Ω (-)
B	30	3.09	B	30	3.09	B	30	3.09
E1	240	5.09	E2	120	3.90	E1	240	5.09
A1	600	2.00	A1	600	2.00	A2	780	1.50
E2	120	3.90	E1	240	5.09	E2	120	3.90
A1	600	2.00	A1	600	2.00	A3	1440	1.03

As a result of the combination of the five wave conditions in different orders, the same wave conditions were run from different initial morphologies. The present study focuses on condition E2 (highlighted in bold on Table 1), which was performed three times, each time commencing from a different initial morphology. Hereinafter, we will refer to E2 performed in sequences 1, 2, and 3 as case 1, case 2 and case 3, respectively. Throughout the measuring campaign, the entire beach profile remained active, i.e., previously moved sediment was still accessible by subsequent waves.

2.1. Wave Condition E2

Wave condition E2 is an energetic wave condition, which is typically classified as 'erosive' condition due to its comparably large dimensionless sediment settling velocity ($\Omega = 3.9$). It is a bichromatic wave condition which presents repeatable wave groups. In the RESIST project, repeatable wave conditions were required for the analysis of detailed sediment transport data which have been investigated in accompanying studies (see, for instance, van der Zanden et al. [29]). Previous studies on data from the same wave flume showed that beach profile evolution under bichromatic waves is similar to random waves, in contrast to monochromatic waves [30].

Details of the wave condition and its two components (with indices 1 and 2 for component 1 and 2, respectively) are summarised in Table 2. The target spectral wave height H_s close to the wave paddle was 0.49 m and the mean primary wave period $T_p = 2/(f_1 + f_2)$ was 3.7 s. T_g presents the period of one wave group ($T_g = 1/(f_1 - f_2)$), each comprising three short waves. T_r corresponds to the repetition period, i.e., the period after which a defined number of wave groups repeats exactly. In the present case, $T_r = 2 \cdot T_g$. The waves were fully modulated, i.e., $H_1 = H_2$.

Waves were generated using first-order wave generation without active wave absorption to allow for sufficient stroke length to generate the high energy wave condition. Previous studies, which involved the same [21] and other data [31,32] from the same wave flume, described a minor influence (<1 cm) of basin seiching and other spurious long waves.

Each case of condition E2 consisted of four 30 min runs resulting in a total duration of 120 min of condition E2 in each case.

Table 2. Target values of wave condition E2.

H_1 (m)	f_1 (Hz)	H_2 (m)	f_2 (Hz)	H_s (m)	T_p (s)	T_g (s)	T_r (s)
0.245	0.3041	0.245	0.2365	0.49	3.7	14.8	29.6

2.2. Measurements

The beach profile was measured along a central line of the flume after every 30 min run as well as before the start of condition E2. The profile measurements were performed by means of a mechanical profiler with a spacial resolution of 0.02 m and a vertical measuring accuracy of 0.01 m.

Information on the locations of the runup and rundown limit as well as on the offshore and onshore limit of wave breaking were acquired by means of visual observations during the first 5–10 min of each run. Wave breaking was identified visually as the point where the wave has started overturning (breaking point [33]) and before the collapsing wave hits the water surface (plunge point [34]). Owing to the bichromatic, i.e., repeatable, wave conditions, the breaking limits can be identified well by means of visual observation of a certain number of waves. Hence, the outer breaking location can be associated with breaking of the largest waves, the inner breaking location relates to breaking of the smallest waves of the group. The region offshore of the outer breaking location is hereafter defined as shoaling zone; the region onshore of the outer breaking is referred to as surf zone. The surf zone is further divided into the breaking zone, where most of the waves, large and small, break (region between outer and inner breaking), and the inner surf zone, where waves propagate as broken waves or turbulent bores (region onshore of the inner breaking up to the rundown limit) (adapted from [33,35]).

Water surface elevation was measured by means of resistive wave gauges (RWG) and acoustic wave gauges (AWG) at a frequency of 40 Hz. RWGs were primarily deployed in the deeper part of the flume and the shoaling region while AWGs were primarily deployed in the shoaling, surf and swash zones. In addition to water surface elevation, AWGs also measure exposed bed levels in the swash zone (see Section 3.3). In the inner surf and swash zone, i.e., between $x \approx -2$ to 7 m, AWGs had a high spatial resolution of circa 1 m. The theoretical measuring accuracy of the AWGs is 0.2 mm, except for the two most onshore located AWGs for which it is 0.02 mm [29].

The three-dimensional velocity was measured at five fixed locations by means of acoustic doppler velocimeters (ADV) at a frequency of 100 Hz. The ADVs were deployed at a vertical elevation of 0.03 m above the initial bed and repositioned before the start of each run. An additional ADV was deployed at a mobile frame that is mounted to a mobile trolley allowing the horizontal and vertical positioning of the instrumentation.

In Table 3, the instrument locations at the beginning of the first run of case 1 are summarised. For cases 2 and 3, the instrument locations were highly similar to case 1 with small variations between 2 and 20 cm in cross-shore direction which arise due to small variations in the handmade initial profiles of each sequence. In case 3, an additional ADV was deployed at $x = -4.52$ m at 3 cm above the initial bed.

Table 3. Instrument locations and vertical elevations above the bed at the start of the first run of case 1.

Device	Quantity	*x*-Location (in m) (vertical elevation above the bed (in m), where applicable)
RWG	12	−63.4, −48.22, −46.71, −42.3, −35.23, −31.16, −27.12, −23.18, −19.21, −17.42, −15.66, −11.3
AWG	19	−56.04, −44.94, −21.85, −20.55, −14.66, −13.26, −9.57, −7.38, −5.57, −3.44, −1.57, −0.52, 0.47, 1.25, 2.31, 3.5, 4.55, 5.56, 6.51
ADV	6	−11.39 (0.085), −1.54 (0.03), −0.52 (0.03), 0.27 (0.03), 1.28 (0.03), 2.26 (0.03)

2.3. Initial Conditions

As aforementioned, condition E2 was performed three times but from varying initial morphologies. These morphologies are shown in Figure 1 and a brief description of the main features is given in Table 4.

In case 1, condition E2 was run from a profile that developed under a (slightly) low energy condition with a dimensionless sediment settling velocity $\Omega = 2$. The resulting profile is characterised by a pronounced bar–trough relief (main breaker bar) with a secondary inner bar and a small berm (between $x \approx 2$ to 3 m). This profile configuration can be classified following [9] as an intermediate beach with elements of the 'rhythmic bar and beach state' (RBB), such as a pronounced bar–trough relief and a small berm, even though three-dimensional features cannot be defined in the present study. It should be noted that the low energy condition prior to E2 was run from a high energy profile (see Table 1) after which limited shoreline recovery occured during the low energy condition [21]. Prior to the start of E2, the bar had a considerable volume, and it was located at $x \approx -7.7$ m (Figure 1).

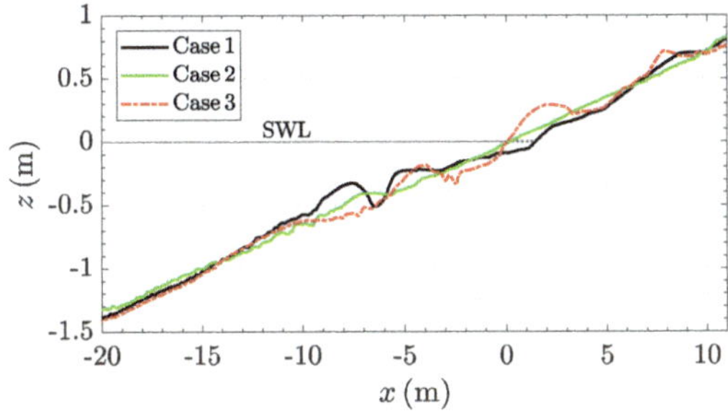

Figure 1. Initial morphologies of the three cases.

In case 2, condition E2 commenced from an almost plane, 1/15 sloped beach profile. Case 2 was run after 30 min of a random wave condition (condition B in Table 1) with $\Omega = 3.09$ (performed to homogenise and compact the manually shaped bed).

In case 3, the initial profile is a low energy beach with elements of a low energy intermediate 'ridge-tunnel type' (RR) and of a reflective beach [9]. Noticeable features of the initial profile are a small bar, a pronounced berm (at $x \approx 0$ to 3 m) and a runnel. The profile had developed under a low energy wave condition with $\Omega = 1.5$ (see Table 1).

Table 4. Initial beach morphologies of the three cases.

Case	Overall Profile	Bar	Swash Berm
Case 1	Profile after slightly accretive condition	Offshore bar	Small swash berm
Case 2	(Almost) plane profile	Barless	No swash berm
Case 3	Low energy beach	Small bar in shallow water	Large berm with runnel

The initial shoreline location is relatively similar for cases 2 and 3 and is further offshore, i.e., the beach is wider, than for case 1. In both cases 1 and 3, the initial shoreline is found near the toe of the swash berm.

The same wave condition was performed in each case. Figure 2 shows a segment of circa 100 s of the water surface elevation measured at $x \approx -63.4$ m (corresponding to circa 11.8 m distance from the wave paddle) which proves the similarity between the generated waves. The mean difference between the generated water surface time series lies between 0.1–1 % of the target H_s, which is in a similar range as the variability between the waves generated in different runs of E2 but within the same sequence.

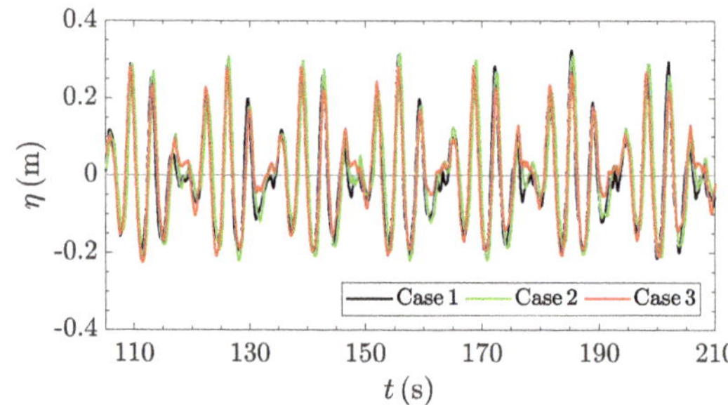

Figure 2. Segment of water surface elevation measurement by RWG at $x \approx -63.4$ m.

3. Data Analysis

3.1. Sediment Transport Calculated from Bed Profile Measurements

Sediment transport rates q (m^2/s) can be calculated at each cross-shore location x_i (m) based on sediment mass conservation ('Exner equation') for any bed level change Δz_i (m) over the associated time interval Δt (s) [30]:

$$q(x_i) = q(x_{i-1}) - \int_{x_{i-1}}^{x_i} (1 - p)\frac{\Delta z}{\Delta t} dx \tag{2}$$

where p (-) presents the porosity of the sediment which was previously measured and which is close to 0.4. q provides direct indication of the direction and magnitude of sediment transported along the profile. This calculation is performed starting from the offshore or onshore boundary (x_{off} and x_{on}, respectively) beyond which no sediment transport occurs (closure limits). Positive values of q indicate onshore sediment transport; negative values correspond to offshore sediment transport.

Integration of q along x and multiplication by the time interval Δt between the profile measurements yields the bulk sediment transport Q (m^3) [30]:

$$Q = \Delta t \int_{x_{off}}^{x_{on}} q(x)dx \tag{3}$$

Q provides an indication of the overall beach response being erosive (negative values of Q) or accretive (positive values of Q).

3.2. Data Treatment

ADVs were primarily placed in the inner surf and swash zone where the sensors are intermittently emerged and submerged. Low quality data, primarily associated with exposed sensors during wave backwash, were detected based on correlation statistics (signal amplitude below 75 counts or signal-to-noise ratio below 15 dB), discarded and not replaced. From the cleaned data, spurious high frequency data are removed by applying a low-pass filter with a cut-off frequency of 3 Hz. The first 60 s and the last 120 s of each ADV time series were discarded.

The repeatable bichromatic wave conditions allow ensemble averaging of hydrodynamic measurements at the repetition period. The detailed procedure to obtain the ensemble averaged data are described in van der Zanden et al. [29]. Subsequently, the maximum wave height H_{max}

is obtained as the difference between the maximum and the minimum water surface elevation of an ensemble.

3.3. Bed Level Changes from AWG Measurements

AWG measurements in the inner surf and swash zone can be used to obtain bed level changes [29,36,37]. AWGs measure the distance to the surface and are calibrated to an initial level, which is, depending on their cross-shore location, either the still water level or the exposed bed. When waves run up and down the beachface between the locations of the rundown and runup limit, the bed is intermittently exposed, which is also measured by the AWGs.

Based on the rundown limit, which was observed during the experimental run, and visual inspection of the AWG time series, the most offshore AWG location where the bed was intermittently exposed and hence, measured by the AWGs, was identified. Because some AWGs have their initial reference with respect to the still water level and to ensure the first short waves have arrived at the beach, the initial bed level is taken as the exposed bed after circa 90 s (after the first three wave groups). Subsequently, a moving minimum approach is applied with a window size corresponding to the repetition period ($T_r = 29.6$ s) to obtain bed level changes between consecutive wave group repeats.

4. Results

4.1. Profile Evolution

The beach profiles obtained by means of the mechanical profiler in intervals of 30 min (after each run) are shown in Figure 3 for cases 1–3 (panels a–c, respectively). The colour changes from light to dark as time progresses. The final profiles of the three cases are shown in Figure 4.

It becomes evident that the beach profile develops towards a similar final profile in the three cases (Figure 4 and black lines in Figure 3), characterised by a main breaker bar–trough system and a secondary bar, a concave profile in the inner surf and swash zone, and the presence of a small berm at the maximum runup location. This final beach configuration is considered to be the equilibrium profile associated with condition E2. The same final beach profile develops, even though the cases start from highly different initial morphologies, indicating that the final profile configuration is largely determined by the wave condition (in line with e.g., [4,7,19,21]).

Largest beach changes occur during the initial stage of condition E2 and beach changes become smaller as the profile approaches the final (equilibrium) configuration (Figure 3). Case 1 (intermediate initial beach profile) is characterised by offshore migration of the breaker bar and of the outer breaking location (from $x \approx -7.7$ m to $x \approx -9.1$ m), including also offshore migration of the secondary inner bar and the inner breaking location. The small existing berm (at $x \approx 2.5$ m) disappears and a new berm evolves close to the maximum runup location. In case 2 (almost plane 1/15 sloped initial profile), a breaker bar develops at $x \approx -8.7$ m and shows only little offshore migration to $x \approx -9.5$ m towards the final beach configuration. The shoreline erodes and a berm forms near the maximum runup location as the beach evolves from a planar to a concave profile. In case 3 (low energy initial profile), the existing bar (at $x \approx -4.4$ m) is eroded and a new bar forms further offshore near the bar location of the final beach configuration (at $x \approx -9.3$ m). The large berm of the initial profile is eroded along with the shoreline. As in cases 1 and 2, a small berm evolves near the maximum runup limit.

A fast evolution of the beach during the initial stage of a wave condition are in line with equilibrium-type beach evolution models, which state that beach changes are faster for a larger disequilibrium of the beach profile for a given wave condition [4,7]. In addition, beach changes in the beginning of relatively energetic wave conditions, such as the present condition E2, were previously reported to occur quickly [21,27].

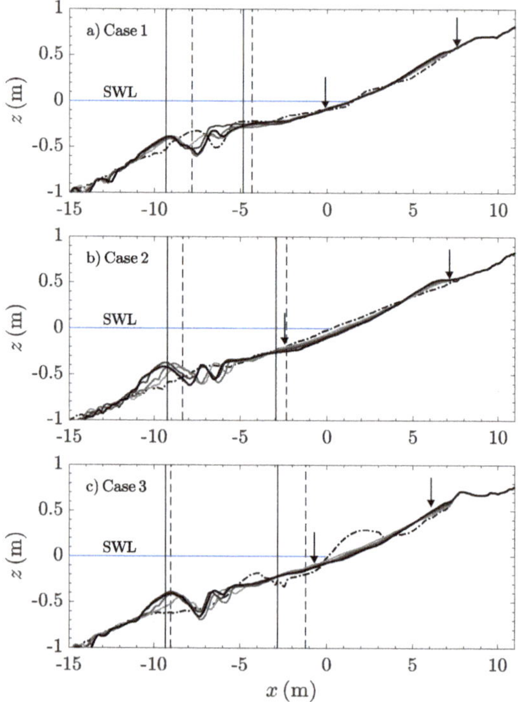

Figure 3. Beach profile evolution in 30 min time intervals. Colour changes from light to dark as time progresses. The dashed-dotted line presents the initial beach profile. Vertical dashed (solid) lines indicate the outer and inner breaking locations taken from observations during the first (last) 30 min run. Arrows indicate the rundown and runup limit during the first 30 min run. (**a**) Case 1; (**b**) Case 2; (**c**) Case 3.

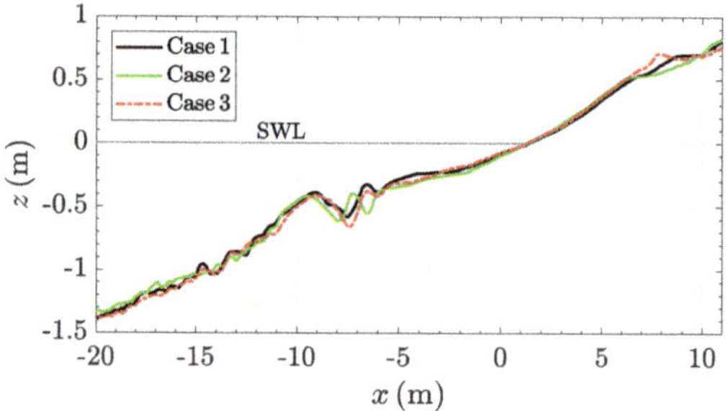

Figure 4. Final morphologies of the three cases.

During the first run, differences in the wave breaking location (vertical dashed lines in Figure 3) highlight the morphodynamic feedback. In case 1, the outer breaking location is controlled by the offshore located bar that is present at the beginning of condition E2. In contrast, in cases 2 and 3, the outer wave breaking location is initially not controlled by the breaker bar. This is because the initial profile is an almost plane profile (case 2) or the bar is small and far onshore so that the largest waves break before arriving at the existing bar location (case 3). In these cases, the outer wave breaking occurs close to the final bar location of condition E2 where the bar is located directly after the first wave run. Generally, the breaker bar is found near the outer breaking location similar to other wave conditions in the same experiment [21] and in other studies, especially for high energy wave conditions [11–13,38,39]. Towards the end of condition E2, the outer wave breaking location (last run indicated by vertical solid lines) is highly similar between the cases. This indicates the importance of the coupling between morphology and hydrodynamics during the first run leading to the equilibrium profile.

Differences between the cases can also be observed in terms of the region between the rundown and runup limits, i.e., the swash excursion. The swash excursion appears to be largest in case 2, which started from the almost plane beach profile. Due to the plane profile, there were no beach features that hindered the waves from going up onto the beach and running back. In case 3, the wave runup goes beyond the berm directly from the beginning of condition E2, which is important for the evolution of the swash berm as will be further shown in Section 4.5.

The evolution of the different initial beach profiles towards the same final (equilibrium) morphology is also supported by the similarity of specific beach parameters, such as the bar and the shoreline location (Figure 5). Figure 5 reveals an equilibrium bar location at $x \approx -9.3$ m and an equilibrium shoreline location at $x \approx 1.3$ m with a variability between the cases of 0.44 m and 0.1 m for the bar and the shoreline, respectively, which can be considered small [13]. The final equilibrium shoreline and bar locations are the same despite the differences in the initial locations. The disequilibrium of the bar and the shoreline in the beginning of condition E2 is largest in case 3 and smallest in case 1.

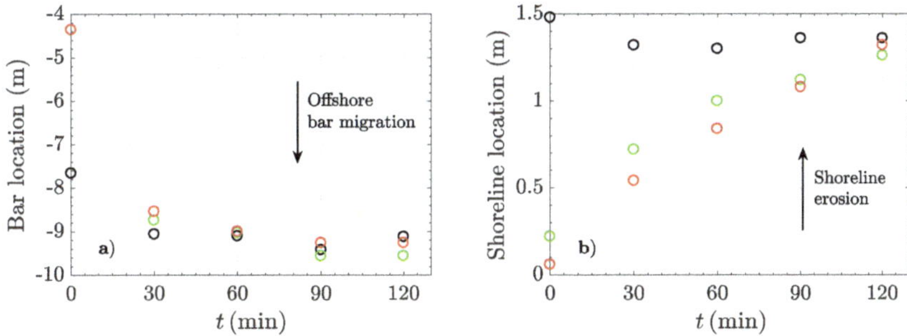

Figure 5. Bar and shoreline location during 120 min of condition E2 for case 1 (○), case 2 (○) and case 3 (○). (**a**) Bar location; (**b**) Shoreline location.

The bar reaches its equilibrium location during the first run of condition E2 (Figure 5a). As aforementioned, the bar of the initial profile of case 3 does not migrate offshore but is eroded and a new bar forms during the first run of E2 in case 3.

Differences in beach evolution between the three cases become also evident for the shoreline (Figure 5b). While the shoreline erodes in cases 2 and 3, it is at a relatively stable location in case 1. This can be related to the initial shoreline in case 1 being close to the equilibrium shoreline location of condition E2. In both case 1 and case 3, the initial shoreline is at the toe of the swash berm. In case 1, this berm is small and hence does not have an important effect on the shoreline location after being eroded, while, in case 3, the berm is large and, hence, strong erosion of the berm leads to a large shift of the shoreline towards the onshore.

It becomes evident that largest morphological changes occur during the first 30 min of condition E2. Therefore, we will primarily focus on this run in what follows.

4.2. Sediment Transport

Sediment transport rates q can be calculated from bed level changes following Equation (2). Figure 6 shows q along the flume during the first run of E2 (solid lines) as well as q over the entire 120 min duration of condition E2 for reference (dashed-dotted lines). Note that q presents a time-averaged quantity over the duration Δt (see Equation (2)) and, therefore, q over the entire 120 min duration can have smaller magnitudes than q during the first run.

Some similitude in the cross-shore shape for both q during the first run as well as for q during the entire duration of E2 can be observed: (1) Slight onshore sediment transport in the shoaling region before wave breaking which is usually associated with wave asymmetries [10,11,40–42]. (2) Offshore sediment transport around the breaking zone, typically associated with offshore directed mean return flow (undertow) [10–12,42]. A wider breaking zone corresponds to a wider region of offshore sediment transport in Figure 6 (compare, for instance, case 1 and case 3). It can be noted that the observed limits of the breaking zone do not always perfectly coincide with the separation between on- and offshore sediment transport which might be related to limitations in the visually observed breaking locations as well as the driving processes acting slightly beyond the observed breaking limits. (3) Positive sediment transport in the most onshore part of the swash zone (i.e., $x > 2$ m) resulting in filling of a previously formed runnel and/or building of a berm [15,18,43]. This berm is smaller and located further onshore than the swash berm of the initial profile (at $x \approx 6$–7 m).

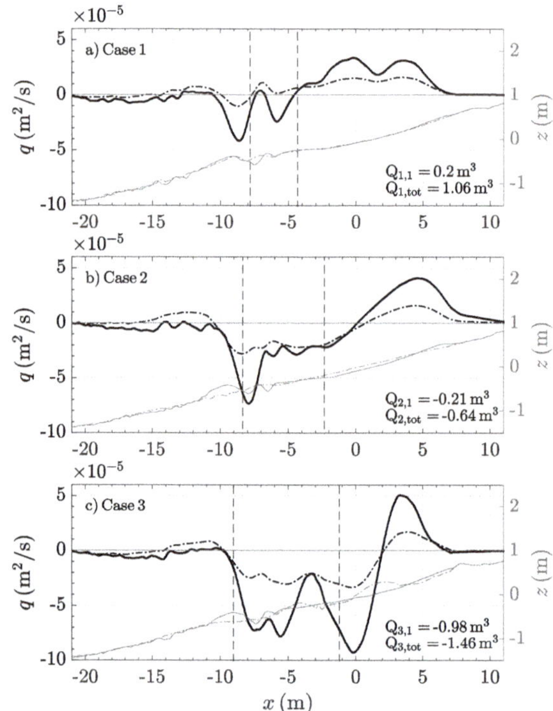

Figure 6. Left ordinate: Sediment transport rate q during the first run (first 30 min) (solid black line) and during the entire duration of E2 (120 min) (dashed-dotted black line). Values of the bulk sediment transport Q are given for the first run (index '1') and for the total duration of condition E2 (index 'tot'). Vertical dashed lines mark the breaking region during the first run. Right ordinate: initial and final profile of each case as dashed-dotted and solid grey lines, respectively. (**a**) Case 1; (**b**) Case 2; (**c**) Case 3.

A closer examination of the three cases reveals differences between the sediment transport rates of the three cases. Maximum offshore sediment tranport magnitudes are larger in cases 2 and 3 compared to case 1 and, as aforementioned, offshore transport occurs over a wider region associated with a wider breaking zone. This can be related to the evolution of the breaker bar: while the size and shape of the bar in case 1 does not change significantly and, hence, does not require additional sediment supply, the bar in cases 2 and 3 forms during the first run of E2. In addition, in case 3, the small initial onshore located bar gets eroded resulting in large sediment transport magnitudes at $x \approx -5\,$m.

In the inner surf and the most offshore region of the swash zone sediment transport magnitudes and directions are highly different between the cases, especially between cases 1 and 3. In case 1, the sediment in the inner surf zone is moved onshore to build the new berm at $x \approx 6\,$m. In contrast, in case 3, large offshore sediment transport magnitudes are present in the inner surf zone which sharply change into onshore directed sediment transport at $x \approx 2\,$m. Combining this with the initial beach profile of case 3, it appears that sediment onshore of the crest of the berm, i.e., $x > 2\,$m, is moved onshore filling up the runnel, while sediment offshore of the berm crest is moved offshore filling up the trough of the initial bar and supplying sediment towards the formation of the newly developing bar.

In terms of sediment exchange between the surf and the swash zone, the above observations indicate that, in case of a low energy initial profile (case 3), a large amount of sediment is drawn from the swash into the surf zone as the berm gets eroded. In contrast, in case of the intermediate energy initial profile (case 1), sediment is moved from the surf into the swash zone.

Overall, the identified differences in sediment transport rates result in different values of the bulk sediment transport Q between the cases. Q_{tot} can also be considered as a quantification of the disequilibrium of the initial beach profile to its final configuration, i.e., how much bulk sediment needs to be moved for the profile to reach its equilibrium for the prevailing wave condition. For the present study, where all cases result in a similar final profile, case 1 (intermediate energy initial profile) requires a positive (onshore) bulk sediment transport to reach the final profile configuration. In contrast, for the initial plane and low energy profile in cases 2 and 3, respectively, negative (offshore), bulk transport is needed to reach the final profile that is more eroded compared to the initial profiles. Negative Q values are largest in case 3 indicating a large disequilibrium of the initial beach morphology.

4.3. Wave Heights along the Flume

The water surface elevation was measured at a high spatial resolution along the flume allowing the investigation of the cross-shore wave height evolution. Figure 7 presents the wave height during the first run of E2 along the flume measured by means of RWGs and AWGs (left ordinate). The wave height is presented as maximum wave height H_{max} of the ensemble averaged data of the first run. H_{max} accounts for all waves in each ensemble and, therefore, presents a useful indicator for the wave height reduction and, hence, energy dissipation of the entire repeatable wave train. The profiles before and after the run as well as the breaking zone are also shown for reference.

A typical cross-shore pattern of wave height evolution can be observed with relatively constant wave heights in the planar bed part of the flume, increase of the wave height due to shoaling when waves get into shallower water and subsequent wave height reduction and associated energy dissipation due to wave breaking. In all three cases, largest values of H_{max} occur at the outer breaking location associated with large instability of the shoaling waves.

Some differences in the cross-shore wave height evolution between the three cases can be observed even though Figure 7 suggests that these differences are relatively small between the cases, especially compared to the large differences in sediment transport rates (Figure 6). Evident differences occur primarily in the breaking zone and can be linked to the differences in the initial profiles. As mentioned in Section 4.1 and as also evident from Figure 7, wave breaking initiates more onshore in case 1 compared to cases 2 and 3 as it is controlled by the initial bar. Moreover, maximum values of H_{max} at the outer breaking location are largest in case 1, which can be associated with the steep offshore slope of the bar promoting wave shoaling. Subsequently, the bar promotes breaking of all, including the smallest, waves resulting in a narrow breaking zone associated with concentrated wave energy dissipation. In this region, a strong mean return flow [10–12] and high suspended sediment concentrations due to breaking-induced turbulence [42] may exist, moving the existing bar offshore . The confined region of wave energy dissipation links up with the narrow region of offshore sediment transport in case 1, as was shown in Figure 6a.

Consequently, the existing bar in case 1 and the associated larger energy dissipation importantly contribute to the protection of the beach from the incoming waves. Waves of large wave height or breaking waves may not get as close onto the beach protecting the beach from the wave impact. This is in line with, for instance, Grasso et al. [24], who reported that a larger bar promoted increased wave energy dissipation resulting in less offshore sediment transport and reduced shore erosion.

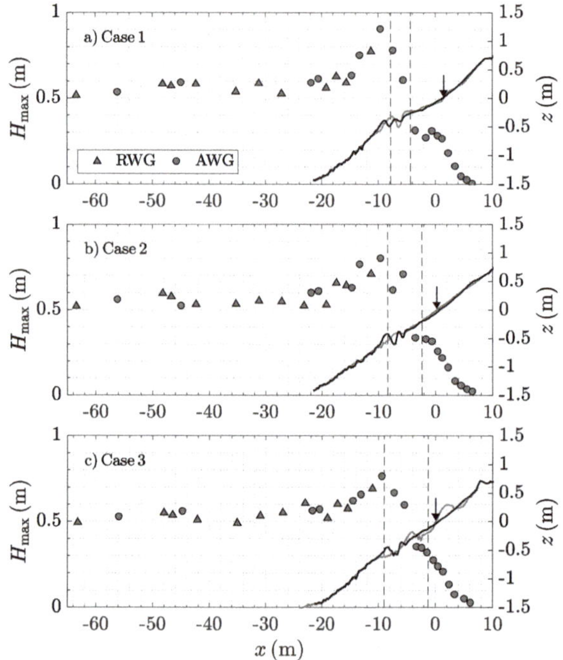

Figure 7. Wave height along the flume during the first run of condition E2. Left ordinate: Maximum wave height obtained from ensemble averaged data. Right ordinate: pre- and post-run profile as grey and black solid line, respectively. Dashed lines indicate the breaking zone (from observations). Arrows indicate the shoreline location of the initial profile. (**a**) Case 1; (**b**) Case 2; (**c**) Case 3.

4.4. Time-Averaged Velocities

Figure 8 shows the time-averaged near-bed flow velocities $\bar{u}$ (Figure 8) in the inner surf and swash zone. For case 2, the velocity measurements could not be included due to limited data quality. From the second run, the mean velocities were found to have a highly similar cross-shore shape and magnitude between the three cases which highlights that a morphodynamic feedback exists, i.e., differences between the mean velocities of cases 1 and 3 during the first run can be linked to the differences in the initial profiles.

A general cross-shore shape of the mean velocities, which are all offshore directed, becomes evident from Figure 8: mean velocities have small magnitudes offshore of the breaking zone which is associated with limited offshore sediment transport (see Figure 6a,c). Note that these velocities were measured at 8 cm in case 1 and at 10 cm in case 3 from the initial bed and hence, velocities measured closer to the bed would be expected to be of even smaller magnitude. Mean velocities become more negative towards the onshore and have their largest magnitude onshore of the breaking zone which may, however, be related to the measuring resolution. Mean velocity magnitudes decrease onshore of $x \approx -1$ m.

The spatial resolution of the velocity measurements is limited in the breaking zone and, hence, does not allow for a clear link between the mean return flow and the breaker bar evolution as well as the investigation of the differences between case 1 and case 3 from the data. In case 1, offshore directed mean velocities would be expected to have a maximum in the breaker zone in line with previous studies [10,11,42]. This is also where largest offshore sediment transport takes place contributing to the offshore migration of the breaker bar (see Figure 6a).

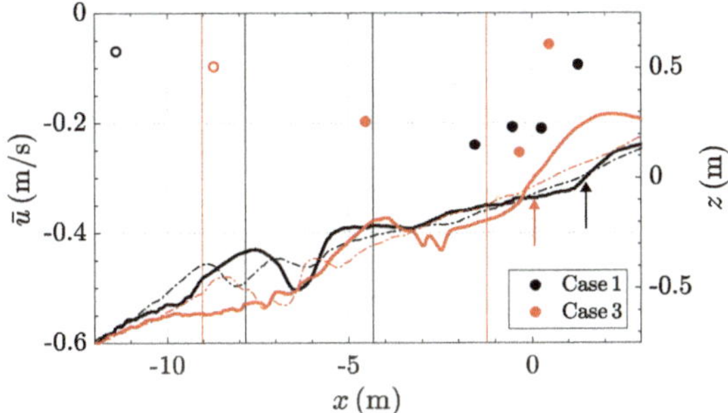

Figure 8. Time-averaged near-bed flow velocities $\bar{u}$ during first run of cases 1 and 3 (left ordinate). Velocites were measured at 3 cm above the initial bed (solid circles), except for the most offshore velocites which were measured at 8 cm and 10 cm in cases 1 and 3, respectively (hollow circles). Right ordinate: pre- and post-run profile measurement as solid and dashed-dotted lines, respectively. Vertical lines indicate the breaking zone during the first run. Arrows indicate the shoreline location of the initial profile.

Around $x \approx 0$ m, i.e., in the inner surf and the beginning of the swash zone, the spatial measuring resolution is larger and Figure 8 indicates differences between the two cases. In case 3, a strong reduction of the mean velocity magnitude occurs over a narrow cross-shore region (two red dots at $x \approx -0.5$ m and 0.5 m). In contrast, in case 1, the decrease of the the velocity magnitude appears to be more gentle and the velocity magnitude is closer to zero at the initial shoreline (see arrows in Figure 8). These differences can be related to differences in the berm between the two cases where the berm is much larger and steeper in case 3. In fact, the reduction of the mean velocity in case 3 occurs almost exactly over the steep offshore-facing berm slope which has similarly been observed in the field [43]. Consequently, the large and steep berm decelerates the mean flow velocity towards its crest and increases it towards the toe. Since this happens over a narrow area due to the large slope of the berm, this change in velocity is associated with a large gradient in sediment transport rate (see Figure 6c) mobilising an important amount of sediment at the toe of the berm where large offshore mean velocities coincide with largest offshore sediment transport magnitudes.

In contrast, in case 1, the berm is much smaller, and the velocity reduces more gently. The velocity magnitude is smallest near the toe of the berm, which corresponds to the location of the initial shoreline of case 1. This might be related to the onshore limit of the breaking zone being further offshore as it does not allow unbroken waves to get close onto the beach compared to case 3. This again highlights the importance of the breaker bar promoting wave breaking and hence wave energy dissipation further offshore. The smaller mean velocity magnitudes at the initial shoreline in case 1 are less likely to promote erosion; in fact, Figure 6a suggests onshore sediment transport in this area. Berm formation and onshore sediment transport in the swash zone have been linked to the advection of sediment from the inner surf and/or lower swash zone, even under conditions with negative mean near-bed velocities [17,18].

4.5. Bed Level Changes during the First Run in the Swash Zone

Bed profile measurements were performed between each wave run and hence provide information on bed level changes with a temporal resolution of 30 min. As shown in Section 4.1, largest morphological changes occur during the first run of condition E2. To investigate the bed evolution within this first run in further detail, the bed profile change in the inner surf and swash zone obtained by means of the AWG measurements (as described in Section 3) is shown in Figure 9 (right panels). For reference, the left panels show the corresponding initial and final profile of the first run obtained by means of the mechanical profiler.

Observations can be made in terms of the timing of the evolution towards a concave swash profile and associated differences in the evolution between the cases. In case 1, both erosion of the berm and onshore deposition of sediment (at $x > 4$ m) start from the beginning of the run. At the initial shoreline location ($x \approx 1.5$ m), the profile change is very small with small accretion towards the end of the run which is in line with the slight shoreline accretion (see Figure 5b). Overall, the volume of the sediment deposited onshore of $x > 4$ m is larger than the eroded volume resulting in overall onshore directed sediment transport in this area (see Figures 6a and 9a,b).

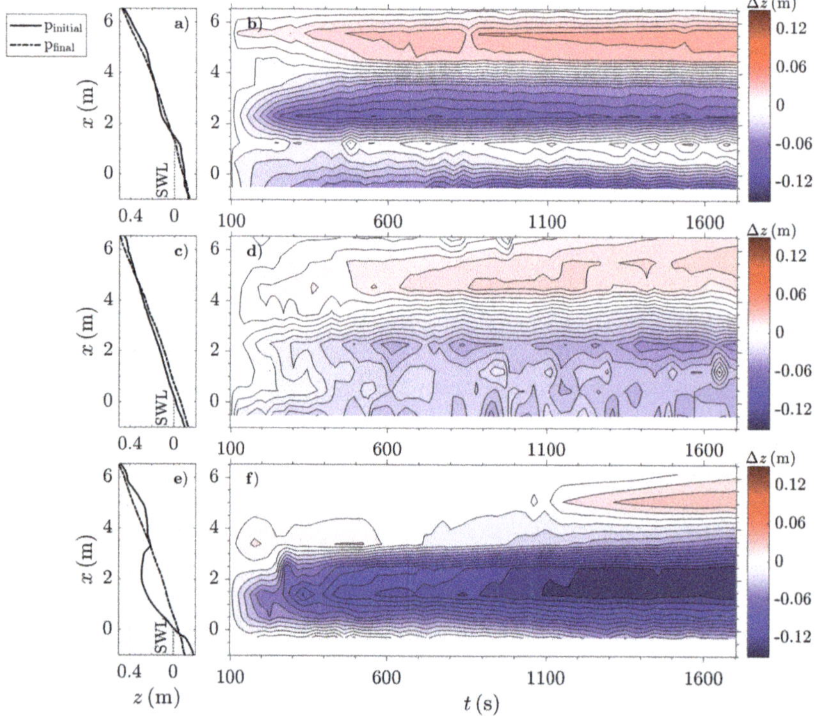

Figure 9. Profile evolution in the inner surf and swash zone within the first 30 min run (between 100 s and 1700 s): top, middle, and bottom panels correspond to case 1, case 2 and case 3, respectively. (**a,c,e**) Profiles measured by mechanical profiler before ($p_{initial}$) and after (p_{final}) the first run of condition E2; (**b,d,f**) Profile change Δz obtained from AWG measurements.

In case 2, which started from the plane profile, no beach features are present and a clear separation (at $x \approx 3.5$ m) between an area of sediment deposition ($x > 3.5$ m) and erosion ($x < 3.5$ m) can be observed towards the evolution of a concave swash profile, which is supported by previous studies [18,29,44]. The net transport of sediment is negative resulting in a net export of sediment from the swash to the inner surf zone (see, for instance, van der Zanden et al. [29]). Sediment deposition initiates almost at the same time as sediment erosion. In both case 1 and case 2, largest morphological changes occur within the first 600 s of condition E2.

Case 3 commenced from the profile with the large berm, which constitutes a more complex case in terms of the temporal evolution of the swash zone in the form of a step-wise evolution. The erosion of the berm starts from the beginning of the run due to the large runup. Initially, overwash over the initial berm transports sediment onshore into the runnel (at $x > 3$ m). The overwash occurs because the wave runup (at least of the largest waves of the groups) from the beginning of the run is sufficiently large to pass the berm (see Figure 3c). The overwash does not directly flow back and hence sediment is deposited. At the same time, erosion at the offshore slope of the berm occurs due to the backwash from which sediment is deposited in the inner surf zone. These processes reduce the height of the berm crest (at $x \approx 1.5$ m), subsequently allowing also smaller waves to pass the berm. This results in further erosion of the berm between circa 600–1000 s of this run without further deposition of sediment onshore. Only when the berm is almost fully eroded, the runnel onshore of the initial berm continues to fill up. This detailed berm destruction by overwash measured in the present laboratory experiments is similar to the beach scarp destruction described under natural conditions by Ruiz de Alegría-Arzaburu et al. [45] and Bonte and Levoy [46].

The described pattern suggests that, in case 3, first, the berm height needs to reduce resulting in a profile that is considered similar to the initial profile of case 1. Subsequently, the evolution pattern is similar to case 1 onshore of $x \approx 4$ m. Consequently, the swash evolution towards the concave (storm) profile in case 3 appears to be temporally delayed compared to case 1 due to the large, initial berm in case 3, which is fully eroded after circa 1000 s.

For all three cases, the bed evolution in the swash zone is the result of a sediment redistribution within the swash and a net exchange of sediment with the inner surf zone. Case 1 leads to net import of sediment into the swash zone, while cases 2 and 3 induce net export of sediment from the swash into the inner surf zone.

5. Conceptual Model and Discussion

The different initial beach morphologies evolve towards the same final equilibrium profile. Differences in the beach evolution are linked to differences in the coupling between beach morphology and hydrodynamics in the three cases. The findings of the present study are, to some extent, also related to the experimental conditions of medium sediment size, a relatively steep beach slope (average slope of 1/15), and moderate wave energy conditions where the same wave condition that started from different initial morphologies generates the same final beach configuration. Other studies, for instance Birrien et al. [20], reported that the initial morphology can also have an effect on the final beach configuration.

It needs to be noted that only the largest differences in hydrodyamics between the three cases could be linked to the initial morphologies because most of the variability in the hydrodynamics between the cases was small, especially compared to the large morphological changes. Analysis of hydrodynamic data at an intrawave scale did not provide evidence of differences between cases to be larger than the variability within one case which highlights limits of the measuring accuracy. Therefore, the analysis focused on quantities averaged over the first run. Time-averaged quantities were found to converge between cases for the subsequent runs highlighting the influence of the initial morphology on the first run. The differences in hydrodynamics between the cases being small is related to the fact that the generated waves are highly similar (Figure 2) and the differences only result from different morphologies and its coupling to the hydrodynamics. An additional measuring limitation is related to the spatial resolution of velocity measurements which were primarily acquired in the region around the shoreline.

Despite these limitations, clear effects of the different initial beach morphologies on equilibrium beach evolution have been identified in the present study. The core findings are summarised in a conceptual model that is shown in Figure 10. Two different initial morphologies are presented in Figure 10a,b, similar to case 1 and case 3, respectively, which were subjected to the same incident wave conditions. The conceptual model has elements of the equilibrium beach state model of

Wright et al. [4] even though the present work does not primarily look into changes between beach states but rather focuses on the processes involved in equilibrium beach evolution.

In the case of an intermediate energy initial profile (Figure 10a), Ω of the incident wave condition is slightly larger than the Ω_0 under which the initial beach profile evolved. The existing bar plays an important role for the equilibrium evolution as it promotes wave breaking and, hence, has a sheltering effect for the beach. Wave breaking at the bar results in fast wave height reduction and associated energy dissipation of all waves around the bar region with offshore sediment transport and largest offshore mean velocity magnitudes at around the onshore limit of the breaking zone. This drives the existing breaker bar offshore towards the equilibrium bar location of the (energetic) wave condition. Changes in the bar shape and volume are limited, and only small additional sediment volumes are drawn from around the breaking zone. Small, broken waves arrive at the shoreline where offshore mean velocities are small and reduce gently over the existing, small berm. As a result, sediment transport is onshore directed in this part of the beach where shoreline changes are limited and the sediment of the small swash berm is moved onshore to form a concave, featureless swash profile.

In the case of a lower energy initial profile (Figure 10b), the wave energy disequilibrium is large ($\Omega \gg \Omega_0$), promoting fast beach evolution towards a more dissipative state [4,7]. The existing breaker bar in shallow water becomes inactive when the higher energy wave condition starts. Outer breaking occurs at the location where the new breaker bar forms, i.e., at the new equilibrium bar location. The smaller waves of the group break close to the shoreline resulting in a wider breaking zone and associated energy dissipation. This promotes offshore sediment transport in the wider breaking zone which contributes to the generation of the new breaker bar further offshore. The swash berm is an important feature of the low energy beach. Mean flow velocities are small at the berm crest and the flow accelerates towards the toe of the berm, promoting erosion close to the initial shoreline. Initial overwash presents an important process that initiates the decay of the berm.

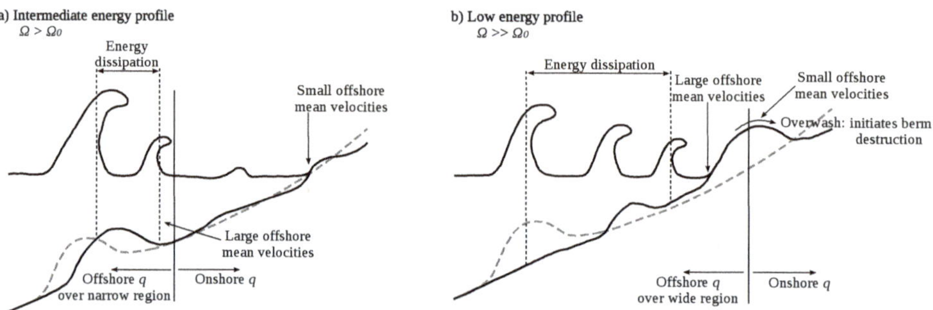

Figure 10. Conceptual model summarising morphodynamics of equilibrium profile evolution for the same incident wave condition and two significantly different initial morphologies. (**a**) Intermediate energy profile; (**b**) Low energy profile. Dashed grey lines present the equilibrium profile towards which the beach evolves.

An important aspect of the two distinct ways of beach evolution is that bulk sediment transport is onshore directed in case 1 but offshore directed in case 3 (Figures 6 and 10). This highlights that the same wave condition, which is classified as a high energy condition and which would, therefore, traditionally be expected to erode the beach, can produce either onshore or offshore sediment transport depending on the initial morphology [7,21]. In the present study, sediment is transported onshore in the inner surf and swash zone in case 1. This is related to specific conditions in comparison to case 3 that can promote onshore sediment transport, which are the narrow breaking zone which is confined around the breaker bar, and the weaker and more gently reducing mean offshore velocities in the inner surf and swash zone. Onshore sediment transport despite offshore mean velocities was observed in previous studies which similarly linked it to weak mean currents that are not strong enough to promote offshore transport leading to a dominance of the onshore-directed wave-related transport [43].

The present study also highlights the importance of the bar and the berm as beach features, each of them functioning in a way to protect the beach against the wave impact. An existing, active breaker bar promotes wave energy dissipation and migrates offshore. This does not require much additional sediment from the inner surf and swash zone, which limits morphological changes in this region.

In case of a low energy profile, the berm presents an important protection against storm events as it presents a large sediment volume that is eroded before further sediment is drawn from the inner surf and swash zone. Previous studies similarly stated that a more recovered beach has a larger volume of sediment that can be eroded initially [7,22,26]. This results in a temporal delay of the inner surf and swash zone to arrive at its more eroded, equilibrium profile. However, the beach is generally considered to evolve fast under energetic wave conditions [21,27,39] and hence the temporal delay has to be considered as a temporally limited protection against wave impact. It would be of interest to investigate further to what extent the size of the berm affects erosion in the inner surf and swash zone. In the present study, initial overwash and resulting onshore sediment transport initiate the destruction of the berm. In case of a larger berm, the run-up of the present wave condition may not go beyond the berm from the beginning of the condition and hence initial berm erosion would be limited to the offshore side of the berm. Other processes, such as strong erosion of the offshore side of the berm or infiltration, may be of importance.

6. Conclusions

The same wave condition was run from three different initial beach morphologies in a large-scale wave flume and resulted in a similar quasi-equilibrium beach profile. The three initial morphologies were an intermediate energy profile, a plane profile and a low energy profile. Beach profile and hydrodynamic measurements, as well as calculations of sediment transport and bed level changes, were studied to explore the role of differences in the initial morphology on beach profile evolution towards equilibrium.

The results lead to the following conclusions:

- The beach evolves towards the same final (equilibrium) morphology that is determined by the wave condition. Different initial beach morphologies do not alter this equilibrium beach morphology but produce different sediment transport patterns, associated with differences in the hydrodynamics, to reach the equilibrium morphology.
- Differences in the profile morphology and hydrodynamics are largest during the first 30 min run, highlighting an important coupling between the beach morphology and the hydrodynamics. When the beach profile is more stable (from the second wave run), the hydrodynamic differences that arise from the different beach morphologies, diminish and more detailed measurements would be needed to investigate effects of morphodynamic coupling on equilibrium beach evolution on a more detailed scale.
- The bar of the initial profile—and, more specifically, its size and location—presents an important morphological feature for the equilibrium beach evolution. The size and location of the bar determine if the bar controls the wave breaking location and associated wave energy dissipation. In case of an intermediate energy profile with an existing breaker bar, wave energy dissipation is confined around the bar in a narrow zone leading to offshore sediment transport in this region and providing a sheltering effect for the inner surf and swash zone. In case of a low energy profile, the initial bar does not control the wave breaking location and wave energy dissipation occurs over a wider cross-shore region associated with offshore sediment transport.
- The large swash berm of the initial low energy profile plays an important role for the equilibrium beach evolution. The berm presents a large source of sediment which is partly moved onshore to fill up the runnel and partly moved offshore towards the breaker bar. Mean velocities are small at

the berm crest with onshore sediment transport rates and accelerate quickly towards the toe of the berm with large offshore sediment transport rates.

- A conceptual model is developed that brings together the most important aspects of the coupling between morphology and hydrodynamics for equilibrium beach evolution for the same incident wave condition but varying initial morphologies. The conceptual model accounts for two importantly different initial morphologies—an intermediate energy and a low energy profile—and accounts for various aspects, including the breaker bar, wave energy dissipation, sediment transport, mean flow velocities and berm decay.

This study has provided insights into the evolution of different initial beach morphologies towards the same equilibrium profile. Beach changes are highly different for different beach morphologies where the bar and berm play an important role. Therefore, it is vital to account for the initial beach morphology and associated hydrodynamic processes in future coastal management and modelling.

Author Contributions: S.E. performed the data analysis and wrote the paper draft. J.M.A. was the leader of the project, the experimental work and the design of the experiments. S.E., J.v.d.Z. and I.C. collaborated in the experimental design and performed the experiments. J.M.A., J.v.d.Z. and I.C. contributed to the analysis and discussion of the results and revised the paper.

Funding: The experiments described in this work were funded by the European Community's Horizon 2020 Programme through the grant to the budget of the Integrated Infrastructure Initiative HYDRALAB$^+$, Contract No. 654110, and were conducted as part of the transnational access project RESIST. S.E. gratefully acknowledges funding from the Department of Civil and Environmental Engineering, Imperial College London, and J.v.d.Z. gratefully acknowledges funding by NWO-TTW (Contract No. 16130). The Imperial College London Open Access Fund is also gratefully acknowledged.

Acknowledgments: We wish to thank fellow researchers in the RESIST project and the CIEM staff (Joaquim Sospedra, Oscar Galego, Andrea Marzeddu) for their support with the experiments as well as the anonymous reviewers for their comments that helped to improve the manuscript.

Conflicts of Interest: The authors declare no conflict of interest.

Abbreviations

The following abbreviations are used in this manuscript:

RESIST	Influence of Storm Sequencing and Beach Recovery on Sediment Transport and Beach Resilience *(project title)*
RWG	Resistive wave gauge
AWG	Acoustic wave gauge
ADV	Acoustic doppler velocimeter
LBT	Longshore bar–trough *(beach state)*
RBB	Rhythmic bar and beach *(beach state)*
RR	Ridge-runnel *(beach state)*
SWL	Still water level

References

1. Short, A.D. Beaches. In *Handbook of Beach and Shoreface Morphodynamics*; Short, A.D., Ed.; John Wiley & Sons: New York, NY, USA, 1999.
2. Wright, L.D.; Thom, B.G. Coastal depositional landforms: A morphodynamic approach. *Prog. Phys. Geogr.* **1977**, *3*, 412–459. [CrossRef]
3. Zhou, Z.; Coco, G.; Townend, I.; Olabarrieta, M.; van der Wegen, M.; Gong, Z.; D'Alpaos, A.; Gao, S.; Jaffe, B.E.; Gelfenbaum, G.; et al. Is "Morphodynamic Equilibrium" an oxymoron? *Earth-Sci. Rev.* **2017**, *165*, 257–267. [CrossRef]
4. Wright, L.D.; Short, A.D.; Green, M.O. Short-term changes in the morphodynamic states of beaches and surf zones: An empirical predictive model. *Mar. Geol.* **1985**, *62*, 339–364. [CrossRef]
5. Ashton, A.; Murray, A.B.; Arnouldt, O. Formation of coastline features by large-scale instabilities induced by high-angle waves. *Nature* **2001**, *414*, 296–300. [CrossRef] [PubMed]
6. Miller, J.K.; Dean, R.G. A simple new shoreline change model. *Coast. Eng.* **2004**, *51*, 531–556. [CrossRef]

7.	Yates, M.L.; Guza, R.T.; O'Reilly, W.C. Equilibrium shoreline response: Observations and modeling. *J. Geophys. Res. Ocean.* **2009**, *114*, 1–16. [CrossRef]

8.	Davidson, M.A.; Splinter, K.D.; Turner, I.L. A simple equilibrium model for predicting shoreline change. *Coast. Eng.* **2013**, *73*, 191–202. [CrossRef]

9.	Wright, L.D.; Short, A.D. Morphodynamic variability of surf zones and beaches: A synthesis. *Mar. Geol.* **1984**, *56*, 93–118. [CrossRef]

10.	Gallagher, E.L.; Elgar, S.; Guza, R.T. Observations of sand bar evolution on a natural beach. *J. Geophys. Res.* **1998**, *103*, 3203–3215. [CrossRef]

11.	Hoefel, F.; Elgar, S. Wave-induced sediment transport and sandbar migration. *Science* **2003**, *299*, 1885–1887. [CrossRef] [PubMed]

12.	Mariño-Tapia, I.J.; Russell, P.E.; O'Hare, T.J.; Davidson, M.A.; Huntley, D.A. Cross-shore sediment transport on natural beaches and its relation to sandbar migration patterns: 1. Field observations and derivation of a transport parameterization. *J. Geophys. Res. Ocean.* **2007**, *112*. [CrossRef]

13.	Eichentopf, S.; Cáceres, I.; Alsina, J.M. Breaker bar morphodynamics under erosive and accretive wave conditions in large-scale experiments. *Coast. Eng.* **2018**, *138*, 36–48. [CrossRef]

14.	Baldock, T.E.; Birrien, F.; Atkinson, A.; Shimamoto, T.; Wu, S.; Callaghan, D.P.; Nielsen, P. Morphological hysteresis in the evolution of beach profiles under sequences of wave climates—Part 1; observations. *Coast. Eng.* **2017**, *128*, 92–105. [CrossRef]

15.	Weir, F.M.; Hughes, M.G.; Baldock, T.E. Beach face and berm morphodynamics fronting a coastal lagoon. *Geomorphology* **2006**, *82*, 331–346. [CrossRef]

16.	Jackson, N.L.; Masselink, G.; Nordstrom, K.F. The role of bore collapse and local shear stresses on the spatial distribution of sediment load in the uprush of an intermediate-state beach. *Mar. Geol.* **2004**, *203*, 109–118. [CrossRef]

17.	Alsina, J.M.; Baldock, T.E.; Hughes, M.G.; Weir, F.; Sierra, J.P. Sediment transport numerical modelling in the swash zone. In Proceedings of the Fifth International Conference on Coastal Dynamics ('05), Barcelona, Spain, 4–8 April 2005. [CrossRef]

18.	Alsina, J.M.; van der Zanden, J.; Cáceres, I.; Ribberink, J.S. The influence of wave groups and wave-swash interactions on sediment transport and bed evolution in the swash zone. *Coast. Eng.* **2018**, *140*, 23–42. [CrossRef]

19.	Grasso, F.; Michallet, H.; Barthélemy, E.; Certain, R. Physical modeling of intermediate cross-shore beach morphology: Transients and equilibrium states. *J. Geophys. Res. Ocean.* **2009**, *114*. [CrossRef]

20.	Birrien, F.; Atkinson, A.; Shimamoto, T.; Baldock, T.E. Hysteresis in the evolution of beach profile parameters under sequences of wave climates—Part 2; Modelling. *Coast. Eng.* **2018**, *133*, 13–25. [CrossRef]

21.	Eichentopf, S.; van der Zanden, J.; Cáceres, I.; Baldock, T.E.; Alsina, J.M. Influence of storm sequencing on breaker bar and shoreline evolution in large-scale experiments. *Coast. Eng.* **2019**, in review.

22.	Scott, T.; Masselink, G.; O'Hare, T.; Saulter, A.; Poate, T.; Russell, P.; Davidson, M.; Conley, D. The extreme 2013/2014 winter storms: Beach recovery along the southwest coast of England. *Mar. Geol.* **2016**, *382*, 224–241. [CrossRef]

23.	Biausque, M.; Senechal, N. Seasonal morphological response of an open sandy beach to winter wave conditions: The example of Biscarrosse beach, SW France. *Geomorphology* **2019**, *332*, 157–169. [CrossRef]

24.	Grasso, F.; Michallet, H.; Barthélemy, E. Experimental simulation of shoreface nourishments under storm events: A morphological, hydrodynamic, and sediment grain size analysis. *Coast. Eng.* **2011**, *58*, 184–193. [CrossRef]

25.	Eichentopf, S.; Karunarathna, H.; Alsina, J.M. Morphodynamics of sandy beaches under the influence of storm sequences—Current research status and future needs. *Water Sci. Eng.* **2019**, *12*, 221–234. [CrossRef]

26.	Coco, G.; Senechal, N.; Rejas, A.; Bryan, K.R.; Capo, S.; Parisot, J.P.; Brown, J.A.; MacMahan, J.H.M. Beach response to a sequence of extreme storms. *Geomorphology* **2014**, *204*, 493–501. [CrossRef]

27.	Morales-Márquez, V.; Orfila, A.; Simarro, G.; Gómez-Pujol, L.; Álvarez-Ellacuría, A.; Conti, D.; Galán, A.; Osorio, A.F.; Marcos, M. Numerical and remote techniques for operational beach management under storm group forcing. *Nat. Hazards Earth Syst. Sci.* **2018**, *18*, 3211–3223. [CrossRef]

28. Eichentopf, S.; Baldock, T.E.; Cáceres, I.; Hurther, D.; Karunarathna, H.; Postacchini, M.; Ranieri, N.; van der Zanden, J.; Alsina, J.M. Influence of storm sequencing and beach recovery on sediment transport and beach resilience (RESIST). In Proceedings of the HYDRALAB⁺ Joint User Meeting, Bucharest, Romania, 21–25 May 2019. Available online: https://hydralab.eu/assets/Proceedings-Hydralab-Joint-User-Meeting-May-23-Bucharest.pdf (accessed on 1 June 2019).
29. Van der Zanden, J.; Cáceres, I.; Eichentopf, S.; Ribberink, J.S.; van der Werf, J.J.; Alsina, J.M. Sand transport processes and bed level changes induced by two alternating laboratory swash events. *Coast. Eng.* **2019**, *152*, 103519. [CrossRef]
30. Baldock, T.E.; Alsina, J.M.; Caceres, I.; Vicinanza, D.; Contestabile, P.; Power, H.; Sanchez-Arcilla, A. Large-scale experiments on beach profile evolution and surf and swash zone sediment transport induced by long waves, wave groups and random waves. *Coast. Eng.* **2011**, *58*, 214–227. [CrossRef]
31. Alsina, J.M.; Padilla, E.M.; Cáceres, I. Sediment transport and beach profile evolution induced by bi-chromatic wave groups with different group periods. *Coast. Eng.* **2016**, *114*, 325–340. [CrossRef]
32. Van der Zanden, J.; van der A, D.A.; Cáceres, I.; Hurther, D.; McLelland, S.J.; Ribberink, J.S.; O'Donoghue, T. Near-Bed Turbulent Kinetic Energy Budget Under a Large-Scale Plunging Breaking Wave Over a Fixed Bar. *J. Geophys. Res. Ocean.* **2018**, *123*, 1429–1456. [CrossRef]
33. Svendsen, I.A.; Madsen, P.A.; Buhr Hansen, J. Wave characteristics in the surf zone. In Proceedings of the 16th International Conference on Coastal Engineering, Hamburg, Germany, 27 August–3 September 1978; pp. 520–539. [CrossRef]
34. Peregrine, D.H. Breaking waves on beaches. *Annu. Rev. Fluid Mech.* **1983**, *15*, 149–178. [CrossRef]
35. Elfrink, B.; Baldock, T. Hydrodynamics and sediment transport in the swash zone: A review and perspectives. *Coast. Eng.* **2002**, *45*, 149–167. [CrossRef]
36. Turner, I.L.; Russell, P.E.; Butt, T. Measurement of wave-by-wave bed-levels in the swash zone. *Coast. Eng.* **2008**, *55*, 1237–1242. [CrossRef]
37. Alsina, J.M.; Cáceres, I.; Brocchini, M.; Baldock, T.E. An experimental study on sediment transport and bed evolution under different swash zone morphological conditions. *Coast. Eng.* **2012**, *68*, 31–43. [CrossRef]
38. Almar, R.; Castelle, B.; Ruessink, B.G.; Sénéchal, N.; Bonneton, P.; Marieu, V. Two- and three-dimensional double-sandbar system behaviour under intense wave forcing and a meso-macro tidal range. *Cont. Shelf Res.* **2010**, *30*, 781–792. [CrossRef]
39. Pape, L.; Plant, N.G.; Ruessink, B.G. On cross-shore migration and equilibrium states of nearshore sandbars. *J. Geophys. Res. Earth Surf.* **2010**, *115*. [CrossRef]
40. Elgar, S.; Gallagher, E.; Guza, R. Nearshore sandbar migration. *J. Geophys. Res.* **2001**, *106*, 11623–11627. [CrossRef]
41. Fernández-Mora, A.; Calvete, D.; Falqués, A.; de Swart, H.E. Onshore sandbar migration in the surf zone: New insights into the wave induced sediment transport mechanisms. *Geophys. Res. Lett.* **2015**, 2869–2877. [CrossRef]
42. Van der Zanden, J.; van der A, D.A.; Hurther, D.; Cáceres, I.; O'Donoghue, T.; Hulscher, S.J.M.H.; Ribberink, J.S. Bedload and suspended load contributions to breaker bar morphodynamics. *Coast. Eng.* **2017**, *129*, 74–92. [CrossRef]
43. Aagaard, T.; Hughes, M.; Møller-Sørensen, R.; Andersen, S. Hydrodynamics and sediment fluxes across an onshore migrating intertidal bar. *J. Coast. Res.* **2006**, *222*, 247–259. [CrossRef]
44. Van der Zanden, J.; Alsina, J.M.; Cáceres, I.; Buijsrogge, R.H.; Ribberink, J.S. Bed level motions and sheet flow processes in the swash zone: Observations with a new conductivity-based concentration measuring technique (CCM⁺). *Coast. Eng.* **2015**, *105*, 47–65. [CrossRef] [CrossRef]
45. Ruiz de Alegría-Arzaburu, A.; Mariño-Tapia, I.; Silva, R.; Pedrozo-Acuña, A. Post-nourishment beach scarp morphodynamics. *J. Coast. Res.* **2013**, 576–581. [CrossRef]
46. Bonte, Y.; Levoy, F. Field experiments of beach scarp erosion during oblique wave, stormy conditions (Normandy, France). *Geomorphology* **2015**, *236*, 132–147. [CrossRef]

Journal of
Marine Science and Engineering

MDPI

Article

Direct Measurements of Bed Shear Stress under Swash Flows on Steep Laboratory Slopes at Medium to Prototype Scales

Daniel Howe [1], Chris E. Blenkinsopp [1,2,*], Ian L. Turner [1], Tom E. Baldock [3] and Jack A. Puleo [4]

1   Water Research Laboratory, School of Civil and Environmental Engineering, UNSW Sydney, Sydney, NSW 2052, Australia; d.howe@unsw.edu.au (D.H.); ian.turner@unsw.edu.au (I.L.T.)
2   Water, Environment and Infrastructure Resilience Research Unit, Department of Architecture and Civil Engineering, University of Bath, Bath BA2 7AY, UK
3   School of Civil Engineering, University of Queensland, St Lucia, QLD 4072, Australia; t.baldock@uq.edu.au
4   Center for Applied Coastal Research, University of Delaware, Newark, DE 19716, USA; jpuleo@udel.edu
*   Correspondence: c.blenkinsopp@bath.ac.uk

Received: 27 August 2019; Accepted: 4 October 2019; Published: 9 October 2019

Abstract: Robust measurements of bed shear stress under wave runup flows are necessary to inform beachface sediment transport modelling. In this study, direct measurements of swash zone bed shear stress were obtained in medium and prototype-scale laboratory experiments on steep slopes. Peak shear stresses coincided with the arrival of uprush swash fronts and high-resolution measurement of swash surface profiles indicated a consistently seaward sloping swash surface with minimal evidence of a landward sloping swash front. The quadratic stress law was applied to back-calculate time-varying friction factors, which were observed to decrease with increasing Reynolds number on smooth slopes, consistent with theory for steady flows. Additionally, friction factors remained relatively constant throughout the swash cycle (except around flow reversal), with a variation of approximately ±20% from the mean value and with only small differences between uprush and backwash. Measured friction factors were observed to be larger than expected when plotted on the Moody or wave friction diagram for a given Reynolds number and relative roughness, consistent with previous field and laboratory studies at various scales.

Keywords: swash; runup; bed shear stress; friction coefficient; shear plate

---

1. Introduction

When a fluid flows past a solid boundary, shear stresses are generated. For rough, fully developed turbulent flows, bed shear stress, τ_0, is commonly related to a friction factor, f, through the quadratic stress law, adapted from the Rayleigh (1876) drag equation [1]:

$$\tau_0 = \frac{1}{2}\rho f|U|U \tag{1}$$

where ρ is fluid density and U is the depth-averaged flow velocity. This formulation is well established for calculating bed shear stress in steady, uniform flow conditions, where f remains constant (e.g., [2,3]). It has also been widely adopted to describe hydrodynamics in the swash zone, where unsteady conditions prevail.

Previous authors have used a range of direct and indirect measurement techniques to calculate friction factors in swash flows (a summary of selected studies is presented in Table 1 and Figure 1), but the results vary widely between studies (see review [4]), with mean values ranging from $f = 0.001$ [5] to $f = 0.04$ [6]. Furthermore, previous studies have often found f to differ between uprush and backwash, and even to vary continuously through the swash cycle (e.g., [7]).

Table 1. Experimental conditions and instrumentation from selected previous studies.

Study	Roughness (mm)	Slope	Study Type	Primary Measurement Technique
Hughes (1995) [8]	0.3–2.0	1:11–1:7	Field	Capacitance gauge
Cox et al. (2000) [6]	6.3	1:10	Laboratory	Laser Doppler velocimetry
Puleo and Holland (2001) [9]	0.26, 0.35, 0.44	1:12	Field	Video camera
Cowen et al. (2003) [7]	Smooth	1:20	Laboratory	Particle image velocimetry
Conley and Griffin (2004) [5]	Medium sand	Dissipative	Field	Hot film anemometer
Barnes et al. (2009) [10]	0.2, 5.8	1:12, 1:10	Laboratory	Shear plate
Kikkert et al. (2012) [11]	1.3, 5.5, 8.4	1:10	Laboratory	Particle image velocimetry
Puleo et al. (2012) [12]	0.2, 0.25	1:20	Field	Acoustic Doppler profiler
Inch et al. (2015) [13]	0.33	1:40	Field	Acoustic Doppler profiler
Pujara et al. (2015) [14]	Smooth plywood	1:12	Laboratory	Shear plate

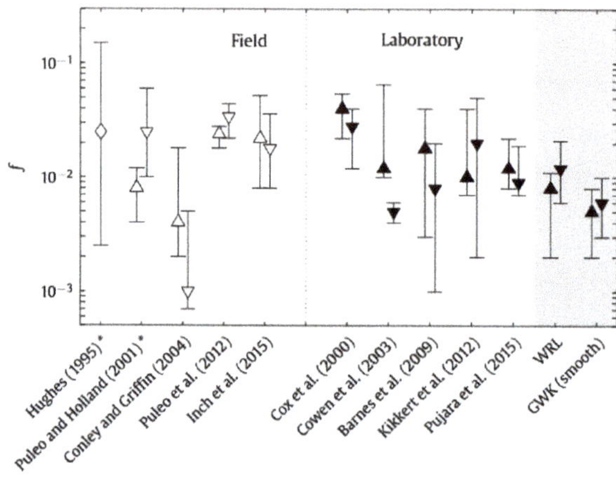

Figure 1. Estimates Darcy friction factor f from current investigation (grey) and selected previous studies of swash flows in fixed-bed laboratory experiments (open symbols) and mobile bed field experiments (filled symbols), showing variation between uprush ($\triangle$/$\triangledown$), and backwash ($\blacktriangle$/$\blacktriangledown$), where reported. * indicates studies where f was calculated using observations of the leading/trailing edge of the swash, rather than the internal flow.

If swash flows are approximated as steady uniform flows in a channel of infinite width, a friction factor can be calculated by applying the Colebrook–White equation [15]:

$$\frac{1}{\sqrt{\lambda}} = -2\log_{10}\left(\frac{k_s}{3.7D_H} + \frac{2.51}{Re\sqrt{\lambda}}\right) \tag{2}$$

where λ is the Darcy friction factor [16], related to the Fanning) friction factor [17], f, by $\lambda = 4f$, while k_s is sand grain roughness, D_H is hydraulic diameter ($D_H = 2h$ for wide rectangular channels, where h is depth), and Re is Reynolds number [18], given by:

$$Re = \frac{uD}{\nu} \tag{3}$$

where u is a characteristic fluid velocity, and ν is kinematic viscosity ($\nu = 10^{-6}$ m^2/s for water).

Numerical models of the nearshore environment typically assume f remains constant through the swash cycle (e.g., [19]). That assumption was tested in this investigation, by back-calculating values of f via the quadratic stress law, using depth-averaged velocity and direct measurements of bed shear stress (provided with a flush-mounted shear plate).

2. Materials and Methods

The results discussed below were obtained through two laboratory experiments in medium and prototype-scale facilities. Both experiments were performed on steep, impermeable slopes. All runup observations from both experiments are reported in bed-parallel coordinates.

2.1. UNSW Water Research Laboratory (WRL)—Medium Scale

One set of experiments was performed using a wave flume (30 m long and 3 m wide) at the Water Research Laboratory (WRL), a facility of the School of Civil and Environmental Engineering, UNSW Sydney, Australia. A 1:3 plywood slope was constructed approximately 8 m from the piston-type wave paddle on top of an existing flat bathymetry with a 1:10 slope at the toe (Figure 2). The water depth was 0.68 m. Water levels were measured using a capacitance wave gauges, and incident and reflected wave spectra were separated using the Mansard and Funke (1980) method [20]. Swash depths were measured using an array of eight Microsonic mic+35 ultrasonic sensors mounted perpendicular to the bed, and spaced at approximately 0.25 m intervals. Bed shear stress was measured with a flush-mounted shear plate [21], with a displacement sensor accuracy of ≤2%. The instruments were connected to a data acquisition PC sampling at 200 Hz for the shear plate, and 40 Hz for the ultrasonic altimeters.

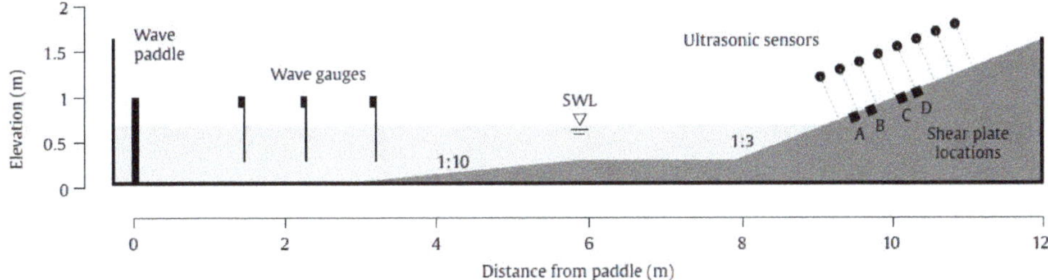

Figure 2. Experimental setup, WRL.

Depth-averaged fluid velocities were calculated using the volume continuity method ([22,23]) based on incremental changes in the measured time-varying swash volume under the ultrasonic sensor array. These continuity-derived velocities had an effective spatial resolution of 0.25 m and a temporal resolution of 200 Hz. Protracted efforts were made to measure swash velocities directly using an electromagnetic current meter mounted close to the bed, but the data were discarded because the instrument did not perform well with the shallow, intermittent flows.

A test program of monochromatic waves was developed, attempting to maximise the range of wave steepness values. Breaker type is generally related to wave steepness using the surf similarity parameter, ξ_0 [24]:

$$\xi_0 = \frac{\tan \beta}{\sqrt{H/L_0}} \tag{4}$$

where β is beach slope, H is wave height, and L_0 is deep water wavelength. Because of the small waves and steep slope used in the WRL experiments, it was more practical to classify the breaker types qualitatively from video records, instead of directly calculating ξ_0.

The test program (Table 2) was repeated four times. The conditions for each repeat were identical, except that the shear plate was installed in a different cross-shore position (A, B, C, and D, in Figure 1) to quantify the spatial variation in bed shear stress. The position of the ultrasonic sensors was modified slightly between each test series, so that two sensors were always located directly above the landward and seaward edge of the shear plate. The most seaward sensor was kept in the same position for all tests to provide a constant reference.

Table 2. Test program, WRL experiments.

Test	Wave Type	T (s)	H (m)	Breaker Type	D_{max} (m)	U_{max} (m/s)	Re_{max}
1	Monochromatic	2.2	0.11	Plunging	0.05	2.0	1.4×10^5
2	Monochromatic	3.2	0.22	Plunging	0.07	2.0	2.3×10^5
3	Monochromatic	3.2	0.16	Collapsing	0.06	1.7	1.3×10^5
4	Monochromatic	5.0	0.16	Collapsing	0.11	2.2	2.1×10^5

2.2. Große Wellenkanal (GWK)—Prototype Scale

A second set of experiments was performed in the Großer Wellenkanal (large wave flume; GWK) at prototype scale. GWK is 310 m long, 5 m wide, and is part of the Forschungzentrum Küste (Coastal Research Centre; FZK) in Hanover, Germany. For this experiment, the instruments were installed on a 1:6 slope, approximately 280 m from the piston-type wave paddle (Figure 3). The water depth was 3.8 m.

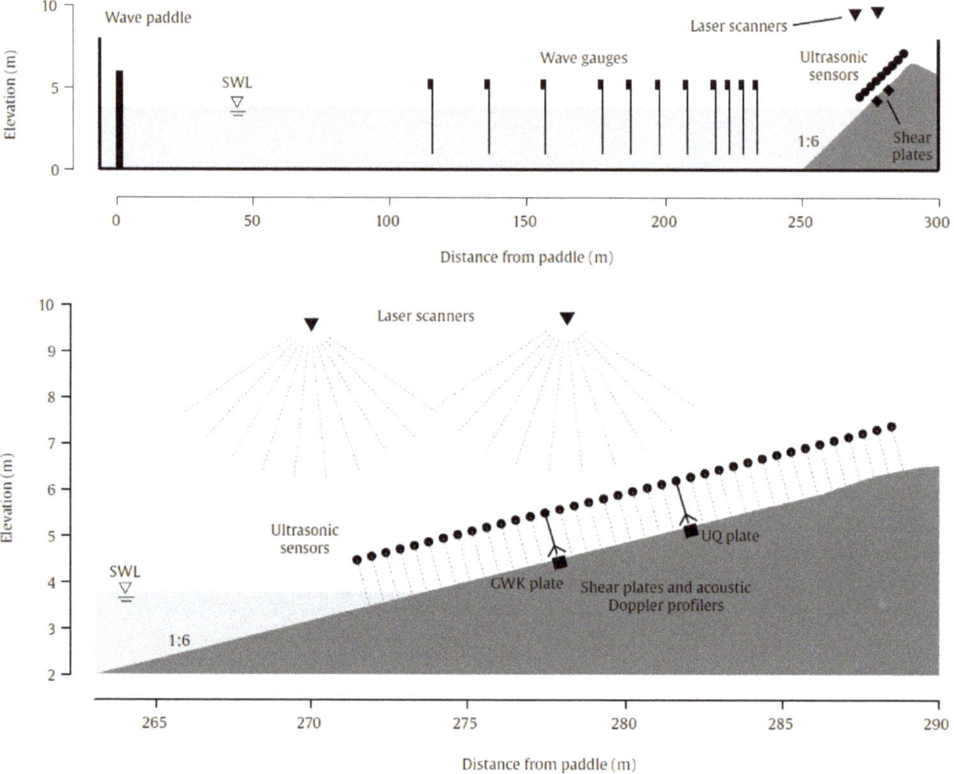

Figure 3. (Upper) Experimental setup, Große Wellenkanal (GWK); (lower) close-up of instrumentation on the subaerial slope.

Wave heights were measured with resistance-type wave gauges sampled at 120 Hz, and instantaneous water depths in the swash zone were provided by an array of 41 Massa M300/95 ultrasonic altimeters sampling at 4 Hz, and mounted to a scaffolding rig at 0.4 m cross-shore intervals at a height of approximately 1 m. The ultrasonic altimeters were supplemented by four laser scanners mounted on two separate crane trolleys suspended above the centre of the flume at distances of 3.5 m offshore, and 4.6 m onshore of the intersection between the still water level and the beach. One SICK

LMS511 and one FARO FOCUS 3D 120S laser scanner were mounted on each trolley, and these instruments were sampled at 35 Hz and 24 Hz, respectively and used to measure time-varying free-surface elevation throughout the inner surf and swash zones (e.g., [25]). Fluid velocities were measured at two cross-shore locations with Vectrino II acoustic Doppler profilers, sampled at 100 Hz.

Depth-averaged fluid velocities were also calculated across the entire swash zone using the volume continuity method from ultrasonic altimeter measurements, yielding an effective spatial resolution of 1 m, and a temporal resolution of 4 Hz. These depth-averaged velocities were validated with the acoustic Doppler profilers, and with laser scanner measurements of the swash leading edge during uprush (Figure 4).

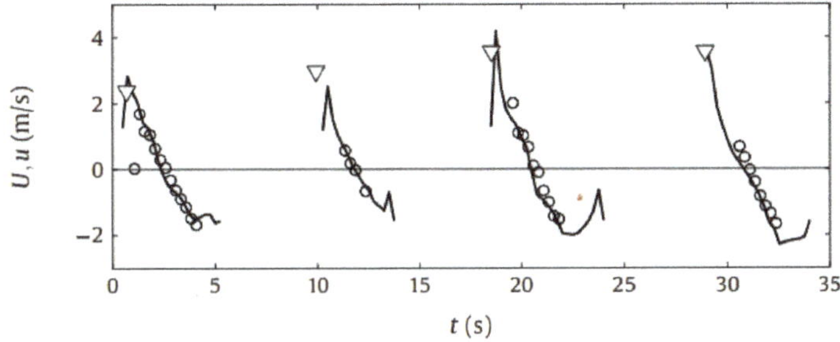

Figure 4. Example comparison of depth-averaged velocities, U derived from acoustic altimeters (black line), an acoustic Doppler profiler 20 mm above the bed (o) and the swash leading edge velocity, u (∇) measured using a laser scanner for four irregular waves during the GWK experiment.

Two shear plates were installed flush with the bed at different cross-shore positions (co-located with the acoustic Doppler profilers), both sampled at 120 Hz (Figure 5). A new bed shear stress transducer (hereafter the GWK shear plate) was used alongside the Barnes and Baldock (2006) [21] instrument (hereafter the University of Queensland (UQ) shear plate). The active face of the GWK shear plate consists of an aluminium disc with a diameter of 150 mm. A flexible rubber gasket sits between the active face and the housing to ensure the shear plate is hermetically sealed (in contrast to the UQ plate). Two dual-beam load cells provide bed shear stress measurements in the x and y directions, with an accuracy of 0.04% [26]. The GWK shear plate and the UQ shear plate were located at distances of 4.4 m and 8.3 m from intersection with the still water level, respectively. A video camera with a wide-angle lens was mounted on the seaward trolley to visually record the swash flows.

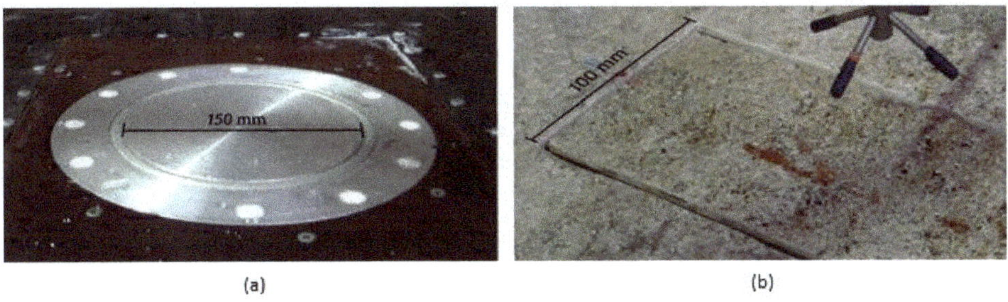

Figure 5. (a) GWK shear plate setup for smooth bed tests; (b) UQ shear plate with moulded rough surface.

A total of 12 monochromatic and irregular (JONSWAP) wave tests were recorded with significant wave height (H_s) between 0.6 m and 1.0 m, and peak wave period (T_p) ranging from 8 s to 14 s (Table 3). The experiments were performed on two different surfaces of contrasting roughness: a 'rough' asphalt surface (tests R1–R2) and a 'smooth' polyethylene surface (tests S1–S9). The surface of each shear plate was carefully prepared to closely match the roughness of the surrounding slope. The sand grain roughness of both shear plates was taken as $k_s = 0.003$ mm for the smooth surface experiments, based on the recommended value for smooth aluminium [27]. During the rough surface experiments, polyurethane moulds of the asphalt surface were glued to the shear plates (Figure 5). Based on photographs and a high-resolution 3D laser survey of the asphalt surface, a roughness value of $k_s = 1.5$ mm was used for the rough surface experiments.

Table 3. Test program, GWK experiments. Note that peak period (T_p) and significant wave height (H_s) are shown for irregular wave cases.

Rough Tests	Smooth Tests	Wave Type	T (s) [1]	H (m) [1]	ξ_0	Breaker Type	D_{max} (m) [2]	U_{max} (m/s) [2]	Re_{max} [2]
R1	S1	Monochromatic	8.0	0.88	1.8	Plunging	0.28	5.4	1.2×10^6
	S2	Monochromatic	8.0	0.99	1.7	Plunging	0.32	4.9	1.3×10^6
	S3	Monochromatic	10.0	0.97	2.1	Plunging	0.33	4.3	1.6×10^6
	S4	Monochromatic	12.0	0.61	3.2	Plunging	0.32	3.8	1.2×10^6
	S5	Monochromatic	12.0	0.72	3.0	Plunging	0.35	5.0	1.7×10^6
R2	S6	Monochromatic	12.0	0.95	2.6	Plunging	0.43	5.6	2.0×10^6
	S7	Monochromatic	14.0	0.63	3.8	Collapsing	0.35	3.7	1.2×10^6
	S8	Monochromatic	14.0	0.85	3.3	Plunging	0.44	4.5	2.1×10^6
R3	S9	Irregular [3]	12.0	0.82	2.8	Plunging	0.28	5.2	9.8×10^5

1. Measured offshore (120 m from paddle).
2. Measured in swash zone (above GWK shear plate).

3. Results

A rich dataset of direct bed shear stress measurements was collected during the WRL and GWK experiments. The shear plates were deployed in multiple locations across the beachface under a wide range of incident wave conditions, allowing bed shear stress to be compared with several other parameters, including bed roughness, experimental scale and cross-shore location.

3.1. Swash Surface Profiles and Boundary Layer Development

Figure 6 illustrates concurrent measurements of the water depth, pressure gradient and bed shear stress, and three snapshots of the water surface profile as a swash front passes the shear plate for monochromatic wave case S2. The water surface profiles are derived from the laser scanner systems positioned above the flume and the data presented are raw data without filtering. The pressure gradient is consistent with that previously observed by Barnes et al. (2009) [10], indicating a water surface dipping seaward, and also documented in field observations [28]. As the swash front passes the shear plate, the well-known immediate spike in the shear stress occurs, but the water surface profile shows little indication of a significant rounded swash front. Thus, there is little evidence that the spike in shear stress and rapid decay is a consequence of, or leads to, the formation of a bull-nose profile at the wave tip, as proposed by [29] on the basis of Eulerian measurements of the flow depth at a point. Indeed, as argued by [30], timeseries measurements of water depth at a fixed point provide little or misleading information about the water surface profile since the swash is unsteady. Consistent with previous data that points to the presence of a well-developed boundary layer at the swash tip (see [30] and [31]), as opposed to a boundary layer growing from the swash tip [29], acoustic Doppler profiler measurements from the GWK experiment show that the boundary layer is at its most fully developed state at the leading edge of the swash, see Figure 7. This is consistent with the boundary layer growing as the flow proceeds up the beach, with the flow at the free surface converging on the front, and with the high shear stress a result of the shallow flow (thin overall boundary layer) and the injection of higher velocity fluid at the wave tip [32].

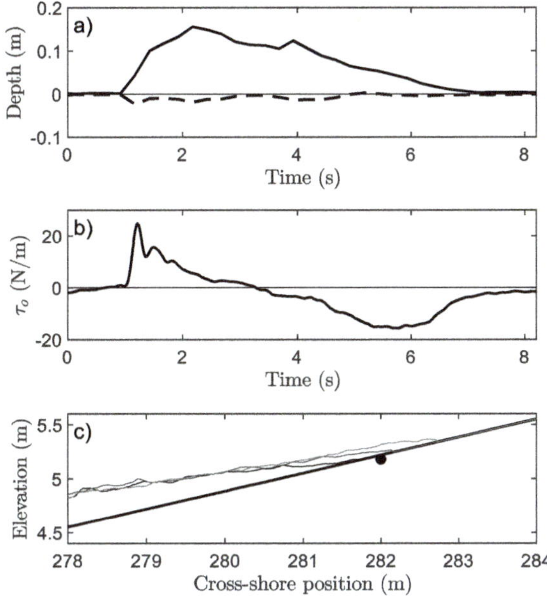

Figure 6. (**a**) Water depth (solid) and pressure gradient (dashed) measured during a single swash event during the GWK experiment for a monochromatic test case S2 ($H = 0.99$ m, $T = 8$ s) (**b**) bed shear stress measured by the UQ shear plate, and (**c**) laser-scanner-derived water surface profiles as the swash front passes the UQ shear plate (black circle) at times 0.8 s (just before arrival at the UQ shear plate), 1.3 s and 1.8 s.

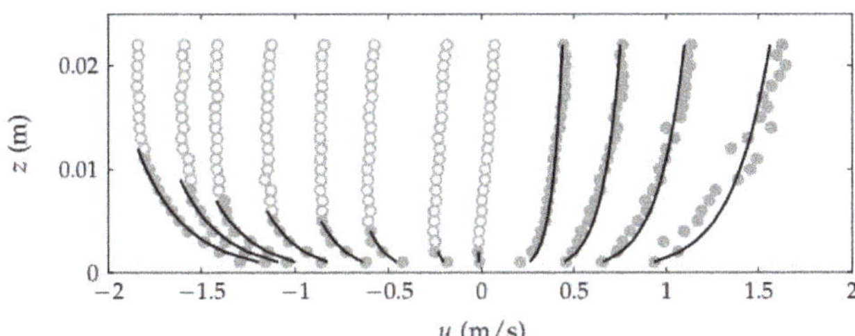

Figure 7. Velocity profiles measured using the acoustic Doppler profiler indicating boundary layer structure through a typical swash cycle for the GWK experiments. Circles show individual velocity measurements and the black lines indicate fitted log-law velocity profiles, filled circles indicate velocity measurements to which the log-law was fitted. Note that the upper limit of the profiler measurements was at $z = 0.022$ m.

3.2. Influence of Bed Roughness on Bed Shear Stress

The GWK experiments were performed on contrasting rough asphalt and smooth polyethylene surfaces. The UQ shear plate was used to measure bed shear stress for the two different surfaces under the same monochromatic wave conditions (Figure 8). As expected, the bed shear stress magnitude over all test cases was larger on the rough surface (by 80% to 100% for uprush and 40% to 60% for backwash). The key feature of larger stresses during uprush is consistent with previous data. The peak backwash bed shear also occurs earlier in the swash cycle for the rough case, compared to the smooth case. This backwash peak is probably caused by the transition to a turbulent boundary layer [32]. The larger peaks on the rough surface may have also been partially caused by the gap (approximately 5 mm wide) between the moulded roughness element attached to the UQ shear plate and the surrounding fixed bed (Figure 5). The gap was kept as small as possible while still allowing free movement of the plate, but may have contributed additional form drag into the bed shear stress measurements. The roughness element interfered with the operation of the GWK shear plate, causing it to produce spurious measurements (which were discarded).

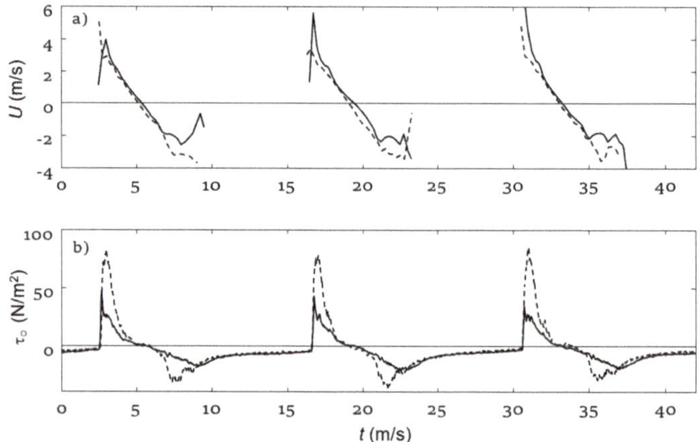

Figure 8. Example comparison of smooth (solid) and rough (dashed) bed measurements for monochromatic waves from the GWK experiment; (**a**) depth-averaged flow velocity; (**b**) bed shear stress measured using the UQ plate.

3.3. Influence of Cross-shore Position on Bed Shear Stress

Cox et al. [6], Barnes et al. [10], and Sumer et al. [33] measured the cross-shore variation in swash zone bed shear stress, and all found τ_o to reach a maximum slightly landward of the still water level, reducing to zero near the maximum uprush limit. Bed shear stress was measured at four different positions on the beachface during the WRL experiments, and the cross-shore bed shear stress envelope (peak value at each location) was calculated for three different monochromatic wave cases (Figure 9). The envelope was normalised with respect to both the peak uprush bed shear, $\tau_{o,maz}$, and the maximum uprush limit, R_x.

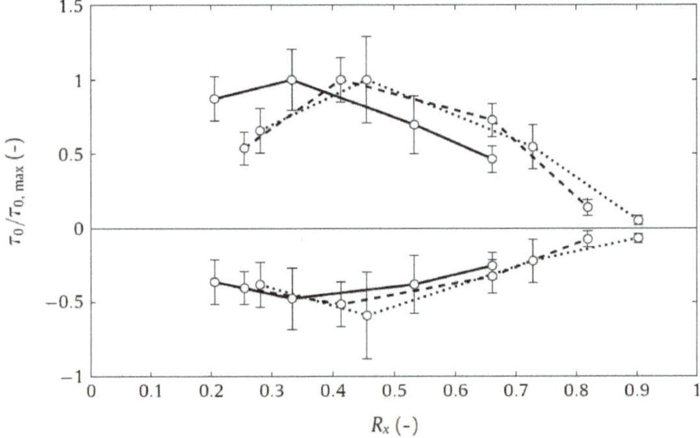

Figure 9. Cross-shore variation of normalised, ensemble-averaged peak bed shear stress envelope from the WRL experiment for three monochromatic test cases: Solid line (H = 0.16 m, T = 5 s), dashed line (H = 0.16 m, T = 3.2 s) and dotted line (H = 0.11 m, T = 2.2 s). Error bars indicate the standard error of the mean.

The WRL bed shear stress measurements showed good agreement with previous studies for the uprush phase, but were proportionally larger during backwash (Figure 10). These backwash bed shear stresses appear larger because they are scaled based on $\tau_{o,max}$ in the uprush phase. The waves in the WRL experiments were generally collapsing. Under these conditions, a relatively large portion of the swash cycle consists of the bore collapse process (where the fluid is accelerating), so $\tau_{o,max}$ occured higher up the beachface. The general factor two difference in the peak stresses between uprush and backwash is explained by the Lagrangian Boundary layer model of Barnes and Baldock [32], whereas the model of Nielsen [34] predicts larger shear stress in the backwash where the backwash is decelerating.

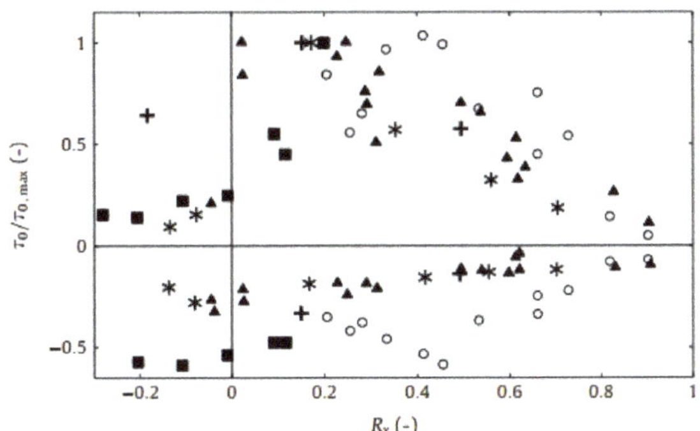

Figure 10. Cross-shore variation of normalised peak bed shear stress from WRL experiments (o), compared with the previous studies of Cox et al. (2000; [6]; +), Barnes et al. (2009; [10], ▲), Sumer et al. (2011, [33]; ■) and Pujara et al. (2015, [14]; ✳).

3.4. Friction Factors

The effective Fanning friction factor for the WRL and GWK experiments was back calculated from time-varying depth-averaged velocity and bed shear stress measurements at each shear plate location, using the quadratic stress law:

$$f(t) = \frac{2\tau(t)}{\rho|U(t)|U(t)} \tag{5}$$

Friction factors calculated in this way tend to approach infinity close to flow reversal, because the divisor is close to zero. For the purposes of this analysis, friction factors were not calculated when velocities were small ($|U| < 0.5$ m/s). Flow reversal occurred rapidly (and the phase discrepancy between τ_o and U was small) because of the steep slopes used in this investigation. Velocities were large. Maximum depth-averaged velocities calculated using the ultrasonic altimeters regularly exceeded 4 m/s in both uprush and backwash during the GWK experiments—larger than velocities reported in previous studies [4].

The period where $|U| < 0.5$ m/s accounted for 20%–25% and 10%–20% of the swash duration for the WRL and GWK experiments, respectively. Bed shear stress was at a minimum in this region: less than 10%–20% of $\tau_{o,max}$ for the WRL experiments, and less than 10%–15% of $\tau_{o,max}$ for the GWK experiments. The time-integrated bed shear stress (i.e., the effective sediment transport potential) around flow reversal was less than 5% of the total.

3.4.1. Friction Factors (WRL Experiment)

The mean value of f was found to be approximately 0.01 for the WRL experiments (Figure 11), and remained fairly constant with time, with a variation of approximately ±40% from the mean value during the backwash phase. This is more consistent than previous studies, some of which present f values on a log scale.

Friction factors were generally highest in the mid-swash zone (Figure 12). There was little variation between uprush and backwash, with the exception of test case 4 (Table 1, H = 0.16 m, T = 5 s), where f was 2–3 times larger during backwash. This is in contrast to some previous studies which have observed significant differences between estimated friction factors during uprush and backwash (Figure 1).

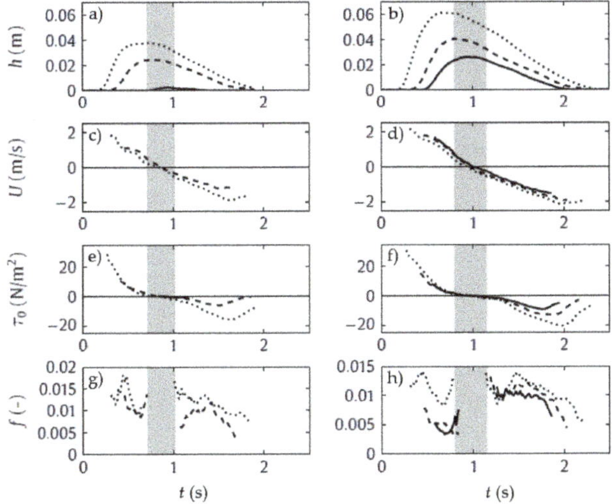

Figure 11. Measurements of monochromatic waves during the WRL experiment. (left column) test case 1 (H = 0.11 m, T = 2.2 s), and (right column) test case 2 (H = 0.22 m, T = 3.2 s) at shear plate positions B (dotted), C (dashed) and D (solid) (see Figure 2). (**a–b**) Flow depth. (**c–d**) Flow velocity. (**e–f**) Bed shear stress. (**g–h**) Time-varying friction factors calculated from ensemble-averaged measurements. The grey region indicates the period over which $|U| < 0.5$ m/s around flow reversal and f was not calculated.

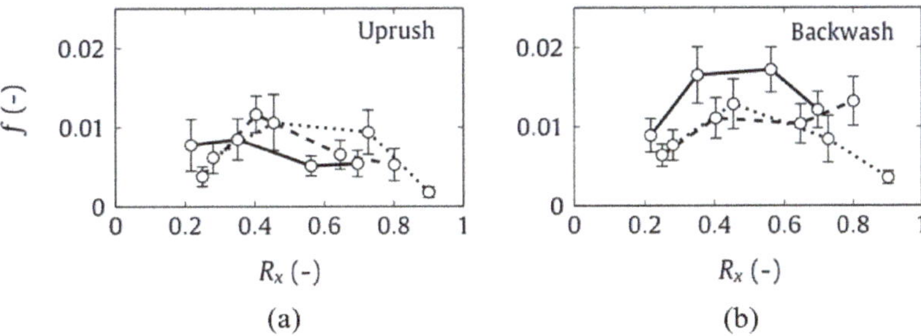

Figure 12. Cross-shore variation of mean friction factor for the WRL experiment for monochromatic test cases 1 (dotted, $H = 0.11$ m, $T = 2.2$ s), 3 (dashed, $H = 0.16$ m, $T = 3.2$ s) and 4 (solid, $H = 0.16$ m, $T = 5$ s) during (**a**) uprush and (**b**) backwash.

3.4.2. Friction Factors (GWK Experiment)

The mean value of f for the GWK experiments was found to be approximately 0.005 for the smooth surface and 0.008 for the rough surface (Figure 13), with 50% of instantaneous values within ±0.0015 and ±0.0025, respectively. On the smooth surface, f remained fairly constant with time for incident waves of varying height (Figure 14), and period (Figure 15), and were observed to be comparable during uprush and backwash. Friction factors were also calculated for a selection of irregular waves from the smooth surface experiments (Figure 16). The friction factors showed more variability within swash cycles than the ensemble-average cases, but the mean value was similar.

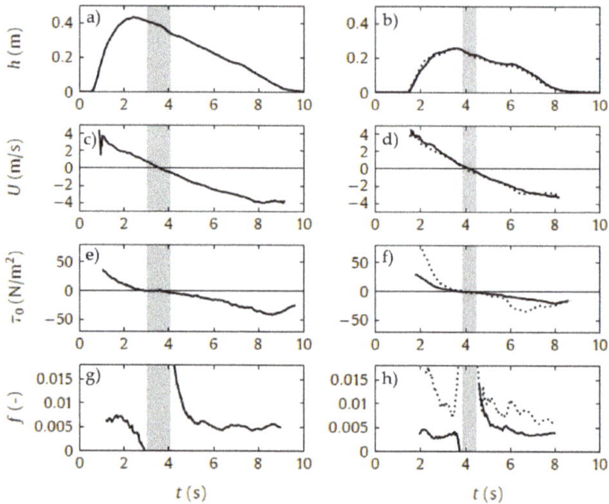

Figure 13. Measurements on a rough and smooth bed from the GWK experiment for a monochromatic test case (H = 0.95 m, T = 12 s; S6 and R2). (left column) shows the smooth bed case in the lower swash at the GWK plate and (right column) presents the results for both the rough (dotted) and smooth (solid) cases in the mid/upper swash measured by the UQ shear plate. (**a–b**) Flow depth. (**c–d**) Flow velocity. (**e–f**) Bed shear stress. (**g–h**) Time-varying friction factors estimated from ensemble-averaged measurements. The grey region indicates the period over which $|U|$ < 0.5 m/s around flow reversal and f was not calculated. Rough bed results from the GWK are not shown due to experimental errors described in Section 3.1.

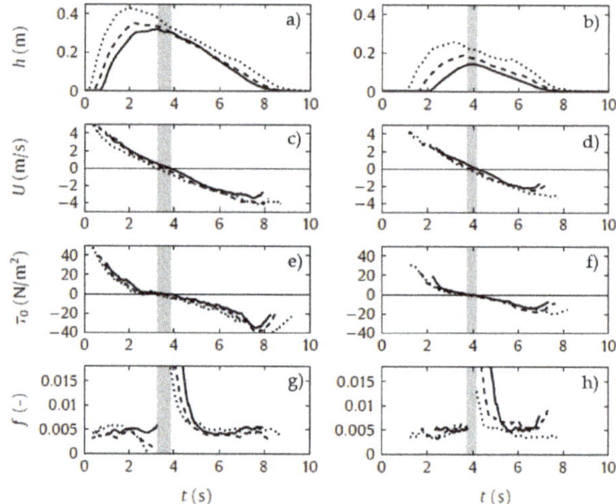

Figure 14. Measurements on a smooth bed from the GWK experiment for monochromatic test cases with constant wave period (T = 12 s) and wave heights of 0.6 m (solid), 0.7 m (dashed) and 0.8 m (dotted). (left column) Shows measurements from the lower swash at the GWK plate, and (right column) presents the results in the mid/upper swash measured by the UQ shear plate. (**a–b**) Flow depth. (**c–d**) Flow velocity. (**e–f**) Bed shear stress. (**g–h**) Time-varying friction factors estimated from ensemble-averaged measurements. The grey region indicates the period over which $|U|$ < 0.5 m/s around flow reversal and f was not calculated.

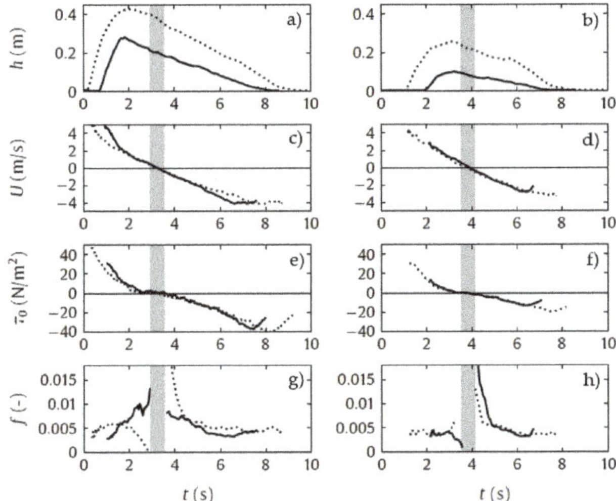

Figure 15. Measurements on a smooth bed from the GWK experiment for monochromatic test cases with constant wave height (H = 0.9 m) and wave periods of 8 s (solid) and 12 s (dotted). (left column) Shows measurements from the lower swash at the GWK plate, and (right column) presents the results in the mid/upper swash measured by the UQ shear plate. (**a–b**) Flow depth. (**c–d**) Flow velocity. (**e–f**) Bed shear stress. (**g–h**) Time-varying friction factors estimated from ensemble-averaged measurements. The grey region indicates the period over which |U| < 0.5 m/s around flow reversal and f was not calculated.

Sediment transport models typically use a constant f for convenience and due to the lack of certainty in the time-varying behavior of this parameter. The calculated friction factors from the GWK and WRL experiments tend to increase around flow reversal, consistent with other studies (e.g., [6,10,14]). This apparent increase in f may be an artefact of division by zero, or incorrect application of (5) because there is no longer fully developed turbulent flow when velocities are small close to flow reversal. If the region close to flow reversal is ignored, f remains relatively constant through the swash cycle in the current experiments. Thus, it appears that the use of a constant value of f is not unreasonable, and because sediment transport models typically use velocity raised to the second or third power [35], any variability in f is likely to have only a secondary effect on calculated transport rates.

3.4.3. Comparison of measured friction factors with previous studies

Typical friction factors for swash uprush from previous work have been collated and plotted on both the Moody diagram (Darcy friction factors) (Figure 16) and the wave friction factor diagram ([36], Figure 17), with the data from this study also added. As expected, the friction factor was observed to decrease with increasing Reynolds number; however, wave friction factors at the same Reynolds numbers (which are defined differently in the two approaches as detailed in the figure captions) are typically 5–10 times larger than Darcy friction factors, which is not completely explained by the factor 4 difference in the definition of f. In both plots, typical friction factors in the uprush are larger than expected for steady uniform flows with the same value of the relative roughness at a given Reynolds number in the current study. This is explained in part by the fact that the uprush flows are unsteady and exhibit a developing boundary layer and flow convergence at the swash tip leading to enhanced bed shear stress. This trend also holds true for all previous studies except point number 3 (Conley and Griffin [5]). Conley and Griffin [5] used a very different measurement approach to all other studies and their results are notably inconsistent with other reported datasets. It is also observed that on the wave friction plot (Figure 17), the result of Cowen et al. [7] plots outside the valid range found by [36] due to the low Reynolds number of the measured flow. For the prototype-scale GWK experiments, it was found that the Colebrook–White equation (Moody diagram) can be used to estimate time-varying friction factors to an accuracy of ±50% if a constant roughness height is applied; however, higher errors (up to approximately 600%) are observed for the other studies.

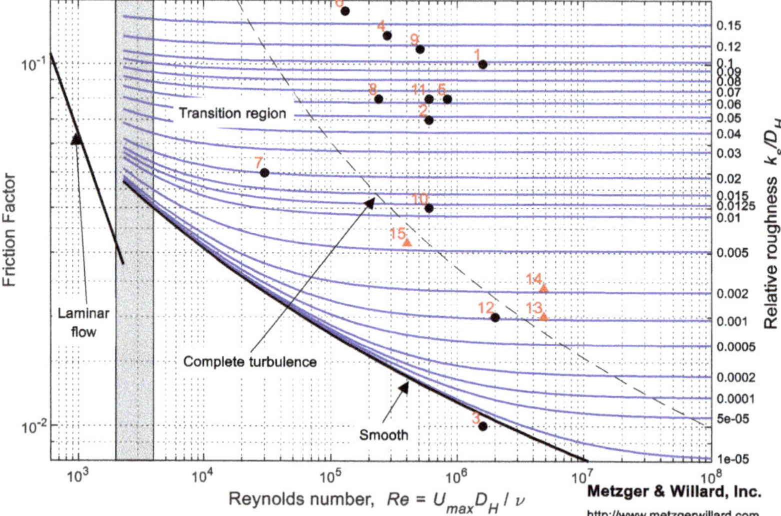

Figure 16. Darcy friction factors (f) plotted on the Moody diagram, using estimated Reynolds number $Re = U_{max}D_H/\nu$, with numbered points corresponding to the studies listed in Table 4. Data from the current experiment is marked with red triangles. Note that data are plotted as $(x, y) = (f, Re)$, and relative roughness is given in Table 4. Moody diagram generated using a modified version of the code developed by [37] with permission of Tom Davis, 2019.

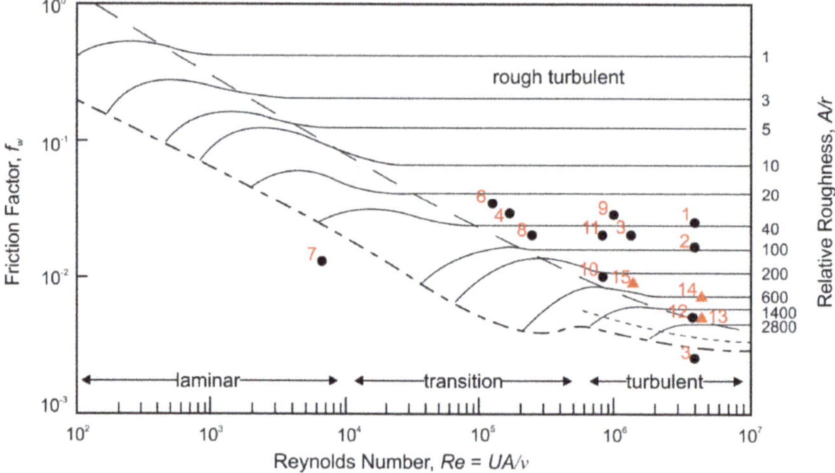

Figure 17. Wave friction factors (f_w) plotted on the Kamphuis wave friction factor diagram [36], using estimated Reynolds number $Re_w = UA/\nu$, with numbered points corresponding to the studies listed in Table 4. Data from the current experiment is marked with red triangles. Note that data are plotted as $(x, y) = (f_w, Re_w)$, and relative roughness is given in Table 4.

Table 4. Typical friction factors for swash uprush from current and previous studies. Darcy friction factors (f) plotted on Moody diagram (Figure 16), using estimated Reynolds number Re = $U_{max}D_H/\nu$. Wave friction factors (f_w) plotted on a Kamphuis wave friction factor diagram, using estimated Reynolds number Re$_w$ = UA/ν. Values for relative roughness are estimated based on typical grain size provided by each author. Backwash friction factors from these studies are typically about half of those measured during uprush, not plotted.

Study No.	Study Author	f (meas)	Re (10^6)	k_s/D_H	f_w (meas)	Re$_w$ (10^6)	A/r
1	Hughes (1995) [8]	0.1	1.6	0.0006	0.025	4	4000
2	Puleo and Holland (2001) [9]	0.07	0.6	0.0011	0.0165	4	5714
3	Conley and Griffin (2004) [5]	0.01	1.6	0.0003	0.0025	4	8000
4	Puleo et al. (2012) [12]	0.12	0.28	0.0006	0.029	0.17	980
5	Inch et al. (2015) [13]	0.08	0.84	0.0005	0.02	1.4	2970
6	Cox et al. (2000) [6]	0.14	0.13	0.0393	0.034	0.13	25
7	Cowen et al. (2003) [7]	0.05	0.03	Smooth	0.013	0.0068	Smooth
8	Barnes et al. (2009) [10]	0.08	0.24	0.0008	0.02	0.25	1250
9	Barnes et al. (2009) [10]	0.11	0.51	0.018	0.028	1.02	110
10	Kikkert et al. (2012) [11]	0.04	0.6	0.0033	0.01	0.84	433
11	Kikkert et al. (2012) [11]	0.08	0.6	0.0210	0.02	0.84	67
12	Pujara et al. (2015) [14]	0.02	2	Smooth	0.005	3.9	Smooth
13	GWK (smooth)	0.02	4.8	0.0001	0.005	4.5	7500
14	GWK (rough)	0.024	4.8	0.0009	0.006	4.5	1000
15	WRL	0.032	0.4	0.0002	0.008	1.4	466

4. Conclusions

This paper presents the first direct, prototype-scale measurements of bed shear stress in the swash zone obtained using a shear plate. Data were collected on both smooth and rough surfaces. Analysis of swash surface profiles confirmed previous observations of a consistently off-shore sloping swash surface except in the few centimetres behind the leading edge of the swash. Arrival of the swash front coincided with the maximum measured shear stress, associated with a well-developed boundary layer at the swash tip. Peak uprush bed shear stresses were found to be typically 2 times greater than during the backwash, with the largest values occurring in the lower/mid swash, landward of the initial bore collapse region. Peak shear stresses were greater for the rough surface by approximately 90% (50%) for uprush (backwash) flows and peak backwash shear stress consistently earlier in the rough bed case, likely caused by an earlier transition to a turbulent boundary layer.

Friction factors were back-calculated using the quadratic stress law, and were found to remain fairly constant (±20% from the mean value) during uprush and backwash, except around flow reversal. This result adds further evidence to support the applicability of the quadratic stress law for swash flows (enabling bed shear stress under wave runup to be approximated by using depth-averaged velocities, and ignoring fluid boundary layers).

Friction factor values were plotted on the Moody and wave friction factor diagram of Kamphuis (1975) along with the results of previous studies. Measured friction factors were observed to decrease with increasing Reynolds numbers as expected for steady, uniform flows. However, it was found that measured friction factors were larger than the values obtained from either plot for a given Reynolds number and relative roughness.

Author Contributions: conceptualization, C.E.B., T.E.B. and I.L.T.; methodology, D.H., C.E.B., I.L.T., T.E.B. and J.A.P.; software, D.H.; validation, D.H.; formal analysis, D.H.; investigation, D.H., C.E.B. and T.E.B.; resources, C.E.B.; data curation, D.H. and C.E.B.; writing—original draft preparation, D.H., C.E.B., I.L.T., T.E.B. and J.A.P.; writing—review and editing, D.H., C.E.B., I.L.T., T.E.B. and J.A.P.; visualization, D.H. and C.E.B.; supervision, C.E.B. and I.L.T.; project administration, C.E.B. and I.L.T.; funding acquisition, C.E.B., I.L.T. and T.E.B.

Funding: This research was funded by the Australian Research Council, grant number DP110101176. J. Puleo was supported by the National Science Foundation (OCE-1332703).

Acknowledgments: The authors would like to acknowledge Michael Allis, Stefan Schimmels, Matthias Kudella and their colleagues at the GWK for their assistance with the experimental work described in this paper.

Conflicts of Interest: The authors declare no conflict of interest.

References

1. Rayleigh, L. On the resistance of fluids. *Lond. Edinb. Dublin Philos. Mag. J. Sci.* **1876**, *2*, 430–441. [CrossRef]
2. Putnam, J.; Johnson, J. The Dissipation of Wave Energy by Bottom Friction. *Trans. Am. Geophys. Union* **1949**, *30*, 67–74. [CrossRef]
3. Jonsson, I.G. Wave boundary layers and friction factors. In Proceedings of the 10th International Conference on Coastal Engineering, Tokyo, Japan, 4–6 September 1966; ASCE: New York, NY, USA, 1966; pp. 127–148.
4. Chardón-Maldonado, P.; Pintado-Patiño, J.C.; Puleo, J.A. Advances in swash-zone research: Small-scale hydrodynamic and sediment transport processes. *Coast. Eng.* **2015**, *115*, 8–25. [CrossRef]
5. Conley, D.C.; Griffin, J.G. Direct measurements of bed stress under swash in the field. *J. Geophys. Res.* **2004**, *109*, C03050. [CrossRef]
6. Cox, D.T.; Hobensack, W.; Sukumaran, A. Bottom Stress in the Inner Surf and Swash Zone. In Proceedings of the 27th International Conference on Coastal Engineering, Sydney, Australia, 16–21 July 2000; ASCE: New York, NY, USA, 2000; pp. 108–119.
7. Cowen, E.A.; Sou, I.M.; Liu, P.L.F.; Raubenheimer, B. Particle Image Velocimetry Measurements within a Laboratory-Generated Swash Zone. *J. Eng. Mech.* **2003**, *129*, 1119–1129. [CrossRef]
8. Hughes, M.G. Friction factors for wave uprush. *J. Coast. Res.* **1995**, *4*, 1089–1098.
9. Puleo, J.A.; Holland, K.T. Estimating swash zone friction coefficients on a sandy beach. *Coast. Eng.* **2001**, *43*, 25–40. [CrossRef]
10. Barnes, M.P.; O'Donoghue, T.; Alsina, J.M.; Baldock, T.E. Direct bed shear stress measurements in bore-driven swash. *Coast. Eng.* **2009**, *56*, 853–867. [CrossRef]
11. Kikkert, G.A.; O'Donoghue, T.; Pokrajac, D.; Dodd, N. Experimental study of bore-driven swash hydrodynamics on impermeable rough slopes. *Coast. Eng.* **2012**, *60*, 149–166. [CrossRef]
12. Puleo, J.A.; Lanckriet, T.M.; Wang, P. Near bed cross-shore velocity profiles, bed shear stress and friction on the foreshore of a microtidal beach. *Coast. Eng.* **2012**, *68*, 6–16. [CrossRef]
13. Inch, K.; Masselink, G.; Puleo, J.A.; Russell, P.E.; Conley, D.C. Vertical structure of near-bed cross-shore flow velocities in the swash zone of a dissipative beach. *Cont. Shelf Res.* **2015**, *101*, 98–108. [CrossRef]
14. Pujara, N.; Liu, P.L.-F.; Yeh, H. The swash of solitary waves on a plane beach: Flow evolution, bed shear stress and run-up. *J. Fluid Mech.* **2015**, *779*, 556–597. [CrossRef]
15. Colebrook, C.F. Turbulent flow in pipes, with particular reference to the transition region between the smooth and rough pipe laws. *J. ICE* **1939**, *11*, 133–156. [CrossRef]
16. Darcy, H. *Recherches Expérimentales Relatives au Mouvement de L'eau Dans les Tuyaux (Experimental Research on the Movement of Water in the Pipes)*; Mallet-Bachelier: Paris, France, 1857.
17. Fanning, J.T. *A Practical Treatise on Hydraulic and Water Supply Engineering*; D. Van Norstrand Company: New York, NY, USA, 1893.
18. Reynolds, O. An experimental investigation of the circumstances which determine whether the motion of water shall be direct or sinuous, and of the law of resistance in parallel channels. *Philos. Trans. R. Soc.* **1883**, *174*, 935–982.
19. Roelvink, D.; Reniers, A.; van Dongeren, A.; van Thiel de Vries, J.; McCall, R.; Lescinski, J. Modelling storm impacts on beaches, dunes and barrier islands. *Coast. Eng.* **2009**, *56*, 1133–1152. [CrossRef]
20. Mansard, E.; Funke, E. The measurement of incident and reflected spectra using a least squares method. In Proceedings of the 17th International Conference on Coastal Engineering, Sydney, Australia, 23–28 March 1980; ASCE: New York, NY, USA, 1980; pp. 154–172.
21. Barnes, M.P.; Baldock, T.E. Bed shear stress measurements in dam break and swash flows. In *International Conference on Civil and Environmental Engineering*; Hiroshima University: Hiroshima, Japan, 2006.
22. Baldock, T.E.; Holmes, P. Swash hydrodynamics on a steep beach. In Proceedings of the Third Coastal Dynamics Conference, Plymouth, UK, 23–27 June 1997.
23. Blenkinsopp, C.E.; Turner, I.L.; Masselink, G.; Russell, P.E. Validation of volume continuity method for estimation of cross-shore swash flow velocity. *Coast. Eng.* **2010**, *57*, 953–958. [CrossRef]

24. Battjes, J.A. Surf similarity. In Proceedings of the 14th International Conference on Coastal Engineering, Copenhagen, Denmark, 24–28 June 1974; ASCE: New York, NY, USA, 1974.

25. Martins, K.; Blenkinsopp, C.E.; Power, H.E.; Bruder, B.; Puleo, J.A.; Bergsma, E. High-resolution monitoring of wave transformation in the surf zone using a LiDAR scanner array. *Coast. Eng.* **2017**, *128*, 37–43. [CrossRef]

26. Pero. *Schubspannungsmesser Für den Einsatz am Meeresgrund (Shear Stress Plate for Use on the Seabed)*; PERO Gesellschaft für Meß und Steuertechnik GmbH: Sickte, Germany, unpublished work; 2007. (In German)

27. Barr, D.I.H. *Tables for the Calculation of Friction in Internal Flows*; Thomas Telford Publishing: London, UK, 1995.

28. Baldock, T.E.; Hughes, M.G. Field observations of instantaneous water slopes and horizontal pressure gradients in the swash zone. *Cont. Shelf Res.* **2006**, *26*, 574–588. [CrossRef]

29. Neilsen, P. Bed shear stress, surface shape and velocity field near the tips of dam-breaks, tsunami and wave runup. *Coast. Eng.* **2018**, *138*, 126–131. [CrossRef]

30. Baldock, T.E. Bed shear stress, surface shape and velocity field near the tips of dam-breaks, tsunami and wave runup. Discussion. *Coast. Eng.* **2018**, *142*, 77–81. [CrossRef]

31. Briganti, R.; Dodd, N.; Pokrajac, D.; O'Donohugh, T. Non linear shallow water modelling of bore-driven swash: Description of the bottom boundary layer. *Coast. Eng.* **2011**, *58*, 463–477. [CrossRef]

32. Barnes, M.P.; Baldock, T.E. A Lagrangian model for boundary layer growth and bed shear stress in the swash zone. *Coast. Eng.* **2010**, *57*, 385–396. [CrossRef]

33. Sumer, B.M.; Sen, M.B.; Karagali, I.; Ceren, B.; Fredsøe, J.; Sottile, M.; Zilioli, L.; Fuhrman, D.R.H. Flow and sediment transport induced by a plunging solitary wave. *J. Geophys. Res.* **2011**, *116*, C01008. [CrossRef]

34. Nielsen, P. Shear stress and sediment transport calculations for swash zone modelling. *Coast. Eng.* **2002**, *45*, 53–60. [CrossRef]

35. Nielsen, P. *Coastal Bottom Boundary Layers and Sediment Transport*; World Scientific: Singapore, 1992.

36. Kamphuis, J.W. Friction factor under oscillatory waves. *J. Waterw. Port C-ASCE* **1975**, *101*, 135–144.

37. Davis, T.G. Moody Diagram. *MATLAB Central File Exchange*. Available online: https://www.mathworks.com/matlabcentral/fileexchange/7747-moody-diagram (accessed on 1 October 2019).

 Journal of
Marine Science and Engineering

Article

Hydrodynamic Effects Produced by Submerged Breakwaters in a Coastal Area with a Curvilinear Shoreline

Francesco Gallerano, Giovanni Cannata * and Federica Palleschi

Department of Civil, Constructional and Environmental Engineering, University of Rome, 001841 Rome, Italy; francesco.gallerano@uniroma1.it (F.G.), federica.palleschi@uniroma1.it (F.P.)
* Correspondence: giovanni.cannata@uniroma1.it; Tel.: +39-064-458-5062

Received: 7 August 2019; Accepted: 22 September 2019; Published: 26 September 2019

Abstract: A three-dimensional numerical study of the hydrodynamic effect produced by a system of submerged breakwaters in a coastal area with a curvilinear shoreline is proposed. The three-dimensional model is based on an integral contravariant formulation of the Navier-Stokes equations in a time-dependent curvilinear coordinate system. The integral form of the contravariant Navier-Stokes equations is numerically integrated by a finite-volume shock-capturing scheme which uses Monotonic Upwind Scheme for Conservation Laws Total Variation Diminishing (MUSCL-TVD) reconstructions and an Harten Lax van Leer Riemann solver (HLL Riemann solver). The numerical model is used to verify whether the presence of a submerged coastal defence structure, in the coastal area with a curvilinear shoreline, is able to modify the wave induced circulation pattern and the hydrodynamic conditions from erosive to accretive.

Keywords: coastal region; hydrodynamics; submerged breakwaters; three-dimensional model; Navier-Stokes equations; integral contravariant formulation; time-dependent curvilinear coordinates

---

1. Introduction

Submerged breakwaters are usually designed to defend the coastline from the erosion induced by breaking waves. The introduction of a system of submerged breakwaters separated by gaps produces two hydrodynamic effects: a local reduction of the water depth, which causes the wave to break earlier in deeper water, and a local variation of the mean water level that induces modifications in the nearshore currents. If correctly designed, the system of submerged breakwaters can induce an accretive circulation pattern, otherwise it can induce an erosive circulation pattern. In the accretive circulation pattern, the wave breaking over the structure produces a wave set-up in the lee of the barrier that is lower than the one in the gaps. Consequently, converging currents occur in the lee of the structure. On the contrary, in the erosive circulation pattern, diverging currents in the lee of the structure take place, because the wave set-up is greater than the one in the gaps in this zone. In the literature, the effectiveness of submerged breakwaters is usually evaluated using the wave transmission coefficient through formulae, i.e., by empirical formulae such as Van der Meer et al. [1], by experimental methods [2–5], or numerical models based on depth integrated motion equations [6–16]. In order to simulate the wave breaking energy dissipation, Tatlock et al. [17] used a vorticity transport equation, while in [15,16] the dispersive terms of the Boussinesq equations are switched off in the surf zone, so that these equations are reduced to nonlinear shallow water equations and are fully solved by shock-capturing schemes. A significant drawback of these models is the necessity to introduce a criterion to establish where and when to switch from one system of equations to the other. Furthermore, it must be underlined that the three-dimensional flow velocity fields cannot be adequately represented by these models. The three-dimensional simulation of wave induced free surface flows can be carried out by numerical models that integrate the three-dimensional Navier-Stokes equations in which the so-called sigma transformation is used. In such a framework, the vertical Cartesian coordinate is transformed in a vertical coordinate that moves with the free surface [18,19]. The adoption of shock-capturing numerical schemes in the σ-coordinate models allows the simulation of breaking waves [20–24]. As opposed to the Boussinesq-type models, no criterion has to be chosen in order to simulate the wave breaking phenomenon. In these σ-coordinates shock-capturing models, the motion equations are written in terms of Cartesian based conserved variables and are solved on a coordinate system that includes a time-varying vertical coordinate. The contravariant formulation of the motion equations allows the integration of these equations on boundary conforming curvilinear grids. In order to simulate hydrodynamic fields and wave breaking on domains that reproduce the complex geometries of the coastal regions, in this work, we adopt an integral contravariant form of the Navier-Stokes equations in a time dependent fully curvilinear coordinate system. The adopted integral contravariant form of the momentum equation is obtained by starting from the momentum time derivative of a fluid material volume and from the Leibniz integral rule for a volume which moves with a velocity that is different from the fluid velocity [25]. The integral contravariant form used in this study has general validity. In fact, as demonstrated in [26], by taking the limit as the volume approaches zero, this integral form is reduced to the complete differential formulation of the contravariant Navier-Stokes equations in a time dependent curvilinear coordinate system, obtained by Luo and Bewley [27]. The adopted integral form of the contravariant Navier-Stokes equations is numerically integrated by a finite-volume shock-capturing scheme which uses Monotonic Upwind Scheme for Conservation Laws Total Variation Diminishing (MUSCL-TVD) reconstructions and an Harten Lax van Leer Riemann solver (HLL Riemann solver). The numerical model obtained is used to simulate, in fully three-dimensional form, the effect on the wave fields and induced nearshore currents produced by the introduction of submerged breakwaters in a coastal area with a curvilinear shoreline. The paper is structured as follows. In Section 2, the integral and contravariant formulation of the motion equations in a system of time-varying curvilinear coordinates is briefly presented. In Section 3, the proposed three-dimensional model is applied to the study of the effects produced by a system of submerged breakwaters separated by gaps in a coastal area with a curvilinear shoreline. Conclusions are drawn in Section 4.

2. Governing Equations

The governing equations adopted in this paper are deduced by the momentum and continuity equations expressed in integral and contravariant form in a time-dependent curvilinear coordinate system proposed in [25].

$$\frac{d}{d\tau}\int_{\Delta V_0}\left(\vec{\tilde{g}}^{(l)}\cdot\vec{g}_{(k)}\rho u^k\sqrt{g}\right)d\xi^1 d\xi^2 d\xi^3 + \sum_{\alpha=1}^{3}\left\{\int_{\Delta A_0^{\alpha+}}\left(\vec{\tilde{g}}^{(l)}\cdot\vec{g}_{(k)}\rho u^k(u^\alpha-v^\alpha)\sqrt{g}\right)d\xi^\beta d\xi^\gamma -\right.$$

$$\left.\int_{\Delta A_0^{\alpha-}}\left(\vec{\tilde{g}}^{(l)}\cdot\vec{g}_{(k)}\rho u^k(u^\alpha-v^\alpha)\sqrt{g}\right)d\xi^\beta d\xi^\gamma\right\} = \int_{\Delta V_0}\left(\vec{\tilde{g}}^{(l)}\cdot\vec{g}_{(k)}\rho f^k\sqrt{g}\right)d\xi^1 d\xi^2 d\xi^3 + \tag{1}$$

$$+\sum_{\alpha=1}^{3}\left\{\int_{\Delta A_0^{\alpha+}}\left(\vec{\tilde{g}}^{(l)}\cdot\vec{g}_{(k)}T^{k\alpha}\sqrt{g}\right)d\xi^\beta d\xi^\gamma - \int_{\Delta A_0^{\alpha-}}\left(\vec{\tilde{g}}^{(l)}\cdot\vec{g}_{(k)}T^{k\alpha}\sqrt{g}\right)d\xi^\beta d\xi^\gamma\right\}$$

$$\frac{d}{d\tau}\int_{\Delta V_0}\left(\rho\sqrt{g}\right)d\xi^1 d\xi^2 d\xi^3 + \sum_{\alpha=1}^{3}\left\{\int_{\Delta A_0^{\alpha+}}\left(\rho(u^\alpha-v^\alpha)\sqrt{g}\right)d\xi^\beta d\xi^\gamma -\right.$$

$$\left.\int_{\Delta A_0^{\alpha-}}\left(\rho(u^\alpha-v^\alpha)\sqrt{g}\right)d\xi^\beta d\xi^\gamma\right\} = 0 \tag{2}$$

where u^k ($k=1,3$) is the contravariant component of the fluid velocity; v^α ($\alpha=1,3$) is the contravariant component of the velocity of the moving coordinate lines; ρ is the water density; f^k and $T^{k\alpha}$ ($k,\alpha=1,3$)

are, respectively, the contravariant component of the external body forces for unit mass vector and the stress tensor. In the above equations τ is the time and ξ^1, ξ^2, ξ^3 are moving curvilinear coordinates obtained from the Cartesian coordinate system (x^1, x^2, x^3, t) by a time-dependent transformation $x^i = x^i(\xi^1, \xi^2, \xi^3, \tau), t = \tau$. Vectors $\vec{g}_{(l)}$ and $\vec{g}^{(l)}$ are, respectively, the covariant and contravariant base vectors of the curvilinear coordinate system; $\sqrt{g}$ is the Jacobian of the transformation [28]. $\Delta V_0 = \Delta\xi^1 \Delta\xi^2 \Delta\xi^3$ is the volume element in the transformed space and $\Delta A_0^{\alpha+}$ and $\Delta A_0^{\alpha-}$ indicate the contour surfaces of the volume ΔV_0 on which ξ^α is constant and which are located at the larger and at the smaller value of ξ^α respectively. Here, the indexes α, β, and γ are cyclic.

Equations (1) and (2) represent the general integral form of the Navier-Stokes equations expressed in a time dependent curvilinear coordinate system. The complete derivation of these equations can be found in [25]. In [26] it has been demonstrated that, by taking the limit as the volume approaches zero, the integral Equations (1) and (2) are reduced to the complete differential form of the contravariant Navier-Stokes equations in a time dependent curvilinear coordinate system that have been proposed in the literature by Luo and Bewley [27].

In this paper, in order to simulate the fully dispersive wave processes and the wave breaking, we adopt the strategy proposed in [26] and obtain the following governing equations

$$
\frac{\partial \overline{Hu^l}}{\partial \tau} = -\frac{1}{\Delta V_0 \sqrt{g_0}} \sum_{\alpha=1}^{3} \left\{ \int_{\Delta A_0^{\alpha+}} \left[\vec{\overline{g}}^{(l)} \cdot \vec{g}_{(k)} Hu^k(u^\alpha - v^\alpha) + \vec{\overline{g}}^{(l)} \cdot \vec{g}^{(\alpha)} GH^2 \right] \sqrt{g_0} d\xi^\beta d\xi^\gamma - \right.
$$

$$
\left. \int_{\Delta A_0^{\alpha-}} \left[\vec{\overline{g}}^{(l)} \cdot \vec{g}_{(k)} Hu^k(u^\alpha - v^\alpha) + \vec{\overline{g}}^{(l)} \cdot \vec{g}^{(\alpha)} GH^2 \right] \sqrt{g_0} d\xi^\beta d\xi^\gamma \right\} +
$$

$$
\frac{1}{\Delta V_0 \sqrt{g_0}} \sum_{\alpha=1}^{3} \left\{ \int_{\Delta A_0^{\alpha+}} \vec{\overline{g}}^{(l)} \cdot \vec{g}^{(\alpha)} GhH \sqrt{g_0} d\xi^\beta d\xi^\gamma - \int_{\Delta A_0^{\alpha-}} \vec{\overline{g}}^{(l)} \cdot \vec{g}^{(\alpha)} GhH \sqrt{g_0} d\xi^\beta d\xi^\gamma \right\} + \qquad (3)
$$

$$
\frac{1}{\Delta V_0 \sqrt{g_0}} \sum_{\alpha=1}^{3} \left\{ \int_{\Delta A_0^{\alpha+}} \vec{\overline{g}}^{(l)} \cdot \vec{g}_{(k)} \frac{T^{k\alpha}}{\rho} H \sqrt{g_0} d\xi^\beta d\xi^\gamma - \int_{\Delta A_0^{\alpha-}} \vec{\overline{g}}^{(l)} \cdot \vec{g}_{(k)} \frac{T^{k\alpha}}{\rho} H \sqrt{g_0} d\xi^\beta d\xi^\gamma \right\}
$$

$$
-\frac{1}{\Delta V_0 \sqrt{g_0}} \int_{\Delta V_0} \vec{\overline{g}}^{(l)} \cdot \vec{g}^{(m)} \frac{\partial q}{\partial \xi^m} H \sqrt{g_0} d\xi^1 d\xi^2 d\xi^3
$$

$$
\frac{\partial \overline{H}}{\partial \tau} = \frac{1}{\Delta A_0^3 \sqrt{g_0}} \sum_{\alpha=1}^{2} \left[\int_0^1 \int_{\Delta \xi_0^{\alpha+}} u^\alpha H \sqrt{g_0} d\xi^\beta d\xi^3 - \int_0^1 \int_{\Delta \xi_0^{\alpha-}} u^\alpha H \sqrt{g_0} d\xi^\beta d\xi^3 \right] \qquad (4)
$$

where $H = h + \eta$ is the total water depth $H = h + \eta$; h is the undisturbed water depth and η is the free surface elevation with respect to the undisturbed water level; G is the gravity acceleration; pressure p is divided into a hydrostatic part, $\rho G(\eta - x^3)$, and a dynamic one, q. The curvilinear coordinates $\xi^1, \xi^2, \xi^3, \tau$ are defined as

$$
\xi^1 = \xi^1(x^1, x^2, x^3); \xi^2 = \xi^2(x^1, x^2, x^3); \xi^3 = \frac{x^3 + h(x^1, x^2)}{H(x^1, x^2, x^3, t)}; \tau = t \qquad (5)
$$

where ξ^1 and ξ^2 are the horizontal boundary conforming curvilinear coordinates and ξ^3 is the time varying vertical coordinate by which the irregular varying domain in the physical space is mapped into a regular fixed domain in the transformed space. $\sqrt{g_0} = \vec{k} \cdot \left| \vec{g}_{(1)} \wedge \vec{g}_{(2)} \right|$, where $\wedge$ indicates the vector product. $\overline{H}$ and $\overline{Hu^l}$ are spatial average values over volume elements defined in the form

$$\overline{H} = \frac{1}{\Delta A_0^3 \sqrt{g_0}} \int_{\Delta A_0^3} H \sqrt{g_0} d\xi^1 d\xi^2$$

$$\overline{Hu^l} = \frac{1}{\Delta V_0 \sqrt{g_0}} \int_{\Delta V_0} \widetilde{\vec{g}}^{(l)} \cdot \vec{g}_{(k)} u^k H \sqrt{g_0} d\xi^1 d\xi^2 d\xi^3$$

(6)

Equations (3) and (4) are numerically integrated by a finite-volume shock-capturing scheme which uses MUSCL-TVD reconstructions and an HLL Riemann solver [25].

3. Results

The three-dimensional model presented has been validated in [24,25]. Here, this model is used to simulate, in fully three-dimensional form, the effect on the wave fields and on the induced nearshore currents produced by the introduction of submerged breakwaters in a coastal area with a curvilinear shoreline.

3.1. Wave Induced Currents in a Coastal Area with a Curvilinear Shoreline

In this subsection we numerically simulate a laboratory experiment [29] of wave propagation and induced nearshore currents in a basin with a curvilinear shoreline. The experiments of Hamm [29] were conducted in a 30 × 30 m basin in which the curvilinear shoreline was obtained by excavating (along the centreline) a rip channel in a plane sloping beach of 1:30 . The basin had an axis of symmetry perpendicular to the wave propagation direction—consequently, in order to save computational time, we numerically reproduce only half of the experimental domain by means of a curvilinear boundary conforming grid that follows the curvilinear shoreline. A plan view of the curvilinear computational grid is shown in Figure 1a, where only one out of every five coordinate lines is visualised. In the same figure, the lines A-A′ and B-B′ are the traces of two cross-sections (one inside the rip channel, $y_{A-A'} = 14.9625$ m, and one at the plane beach, $y_{B-B'} = 1.9875$ m). A three-dimensional view of the bottom of the curvilinear computational grid is shown in Figure 1b. In this test, the input wave conditions are given by a monochromatic wave train with period $T = 1.25$ s and height $H = 0.07$ m.

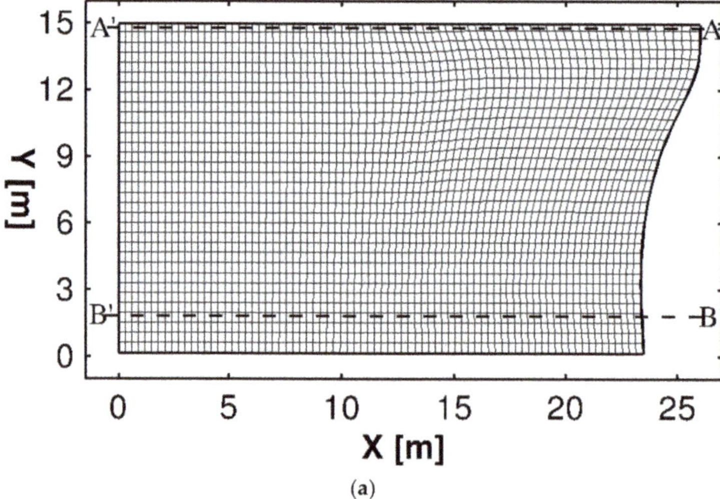

(a)

Figure 1. *Cont.*

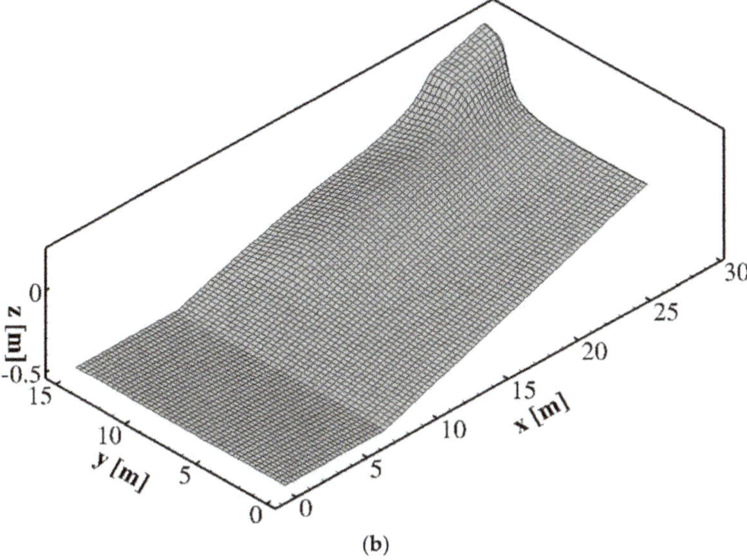

(b)

Figure 1. (**a**): Plan view of the curvilinear computational grid (Only one out of every five coordinate lines is shown). (**b**): three-dimensional view of the bottom.

In Figure 2 we compare, along the two cross-sections, the wave heights obtained by the proposed model with the experimental results of Hamm [29]. From this figure, a good agreement can be noticed between the experimental and numerical results, for both the section placed in the sloping beach and the one placed in the rip channel.

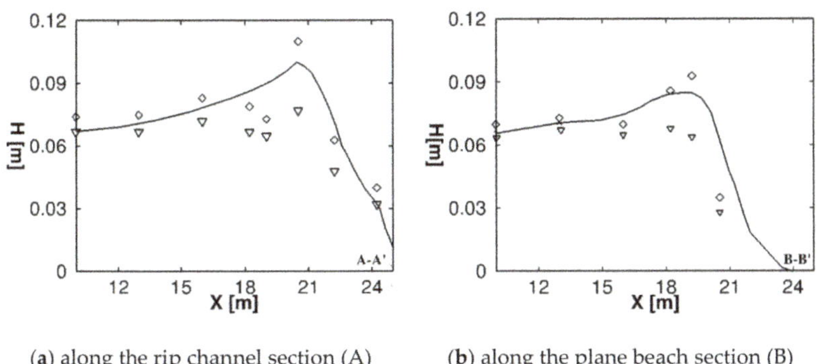

(a) along the rip channel section (A) (b) along the plane beach section (B)

Figure 2. Wave height comparison between the numerical results (solid line) obtained with the proposed model and the experimental data (diamonds represent the experimental data for significant wave height $H_{1/3}$ and gradients represent the experimental data for variance-based wave height $H_\sigma / \sqrt{2}$) from Hamm [29]

The wave induced nearshore current obtained by the proposed model is shown in Figure 3, where only one out of every four time-averaged (over 120 wave periods) flow velocity vectors is visualised. As can be seen in Figure 3, the differences in the wave elevation between the plane beach and rip channel drives an alongshore current that turns offshore producing the rip current at the rip channel position. This circulation pattern can be considered erosive, since it can produce a dragging of the sediments towards the offshore region.

47

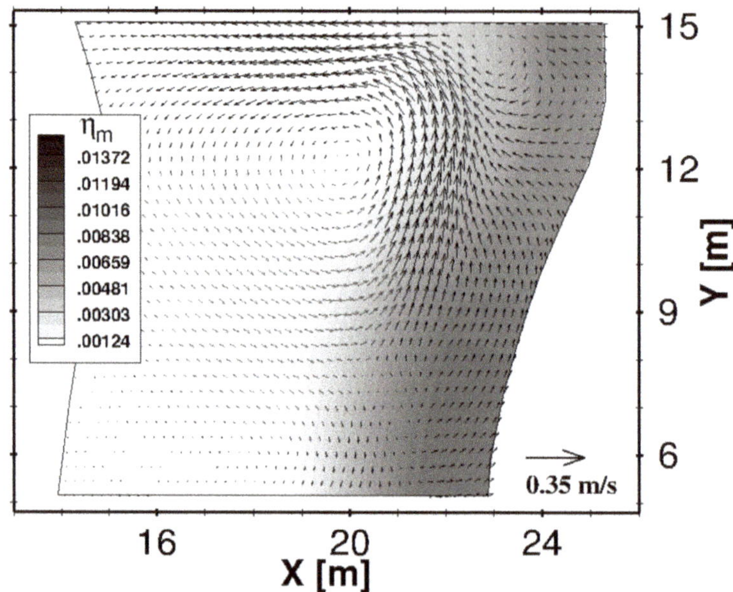

Figure 3. Plan view detail of the time averaged (over 120 wave periods) velocity field. Only one out of every four vectors is shown.

3.2. The Effects of a System of Submerged Breakwaters in a Coastal Area with a Curvilinear Shoreline

The submerged breakwaters are among the most common works realised for the protection of the coastlines affected by erosion phenomena of different intensities. A group of submerged breakwaters separated by gaps protects the coastline against the erosive action of the wave motion as it induces the wave breaking. The energy associated with the incident wave motion is partially reflected offshore and partially dissipated by breaking over the barriers, thus weakening the erosional power of the waves passing over the breakwaters. Furthermore, submerged breakwaters separated by gaps are designed in such a way as to induce, immediately downstream of the breakwaters, gradients of the mean water level, producing circulation patterns that induce bottom accretion (accretive conditions) [30,31]. As experimentally demonstrated [32,33], the distance between the breakwaters and the coastline or the depth of the submerged breakwaters with respect to the undisturbed free surface level are parameters which must be chosen in such a way to favour the development of accretive circulations near the coastline, rather than erosive ones.

Figure 4 shows a schematisation of the hydrodynamic effects produced by a system of submerged breakwaters on the mean water level: Figure 4a refers to a cross-section in correspondence with the barriers; Figure 4b refers to a cross-section along the gap between the barriers. As shows in Figure 4a, close to the submerged breakwaters, due to the wave breaking there is an increase in the mean water level with respect to the still water level (wave set-up η_{1B}). On the onshore side of the breakwater, the wave breaking stops because the waves enter the relatively deeper waters in the lee of the breakwater, then the wave (which is now characterised by a new height) restarts to shoal until it breaks near the coastline. At the coastline, the total set-up with respect to the still water level is given by the sum of the two successive wave set-ups ($\eta_{totB} = \eta_{1B} + \eta_{2B}$). As schematised in Figure 4b, between the barriers, the incident waves propagate without being directly affected by the presence of the barriers and break near the shoreline. Consequently, in the gap between the barriers, the waves set-up η_{1G} is lower than the one (η_{1B}) that takes place over the barriers ($\eta_{1G} < \eta_{1B}$). This difference in the mean water level drives a rip current in the gap that is directed offshore. Furthermore, in the protected area between the barriers and the shoreline, the modifications produced on the incident waves by the presence of the barriers induce nearshore circulations which can be summarised in two different circulation pattern types.

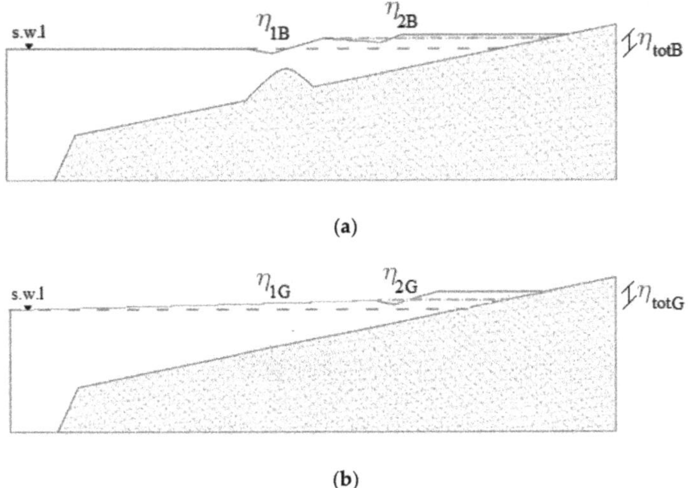

Figure 4. Schematic cross-section and profile of the set-up; (a) section placed over the submerged breakwater (b) section placed in the gap between the barriers.

The first type is characterised by the fact that, near the shoreline, the mean water level in correspondence with the section placed between the barriers is higher than the one in correspondence with the section placed over the barriers ($\eta_{totG} > \eta_{totB}$). This induces a secondary circulation with direction opposite the primary one and thus induces an accumulation of suspended sediments in the sheltered area and the advancement of the shoreline (accretive conditions). This type of circulation pattern is known as four-cell circulation (Figure 5a).

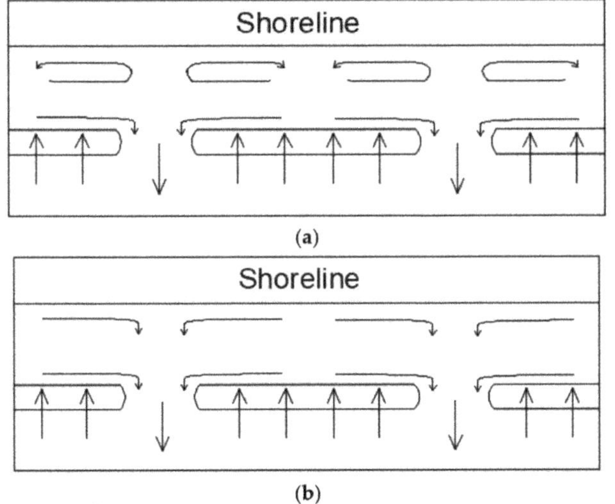

Figure 5. (a) Accretive circulation pattern (b) Erosive circulation pattern.

The second type is characterised by the fact that, near the shoreline, the mean water level in correspondence with the section between the barriers is lower than the one in correspondence with the section placed over the barriers ($\eta_{totG} < \eta_{totB}$). This induces a secondary circulation with the same direction as the primary one. This type of circulation pattern is known as two-cell circulation (Figure 5b). The two-cell circulation produces a dragging of the sediments towards the offshore region, thus favouring the erosion of the seabed at the coastline with possible shoreline regression (erosive conditions).

From a general point of view, in order to obtain accretive conditions, a system of submerged breakwaters separated by gaps has to produce a partial decrease of wave energy and wave height in the lee of the barriers such that the waves transmitted break closer to the shoreline than those at the gaps and with a smaller wave set-up ($\eta_{totG} > \eta_{totB}$). As shown by Ranasinghe et al. [30,31], for given wave characteristics (wave height and period), the main factors of a system of submerged breakwaters separated by gaps that can cause erosive or accretive conditions are: the water depth at the bar crest, c; the ratio between the length of the barrier and the gap width, L_B/L_G; the distance from shoreline to barrier, d. All other factors being equal, as the water depth at the bar crest decreases, the wave height reduction of the transmitted waves increases. It can produce a wave set-up, η_{totB}, lower than the one produced by the waves passing through the gaps, η_{totG} (accretive conditions). Concerning the ratio between the length of the barrier and the gap width, L_B/L_G, all other factors being equal, a reduction of L_G increased the amount of water accumulated in the sheltered area. Consequently, near the shoreline, the wave set-up induced by the transmitted waves η_{totB} can be greater than the one produced by the waves passing through the gaps, η_{totG} (erosive conditions). On the contrary, as the distance from shoreline to barrier increases, all other factors being equal, the distance between the first and second wave breaking enhances, thus reducing the amount of water accumulated in the sheltered area. It can produce a wave set-up, η_{totB}, lower than the one in correspondence of the gaps, η_{totG}, (accretive conditions).

In this section, we verify whether the presence of a submerged coastal defence structure in the coastal area with curvilinear shoreline described in the previous section, is able to modify the wave induced circulation pattern and the hydrodynamic conditions from erosive to accretive. The coastal defence structure is made up of two submerged breakwaters, separated by a gap, similar to the ones used in the experimental test shown in Section 3.1: the barrier length is 3.6 m, the distance between the barriers is 1.8 m, and the average water depth at the bar crest is 2.67 cm. We use the same curvilinear computational grid as in the previous test and numerically simulate two different cases: in Case 1 the submerged breakwaters are positioned inside the surf-zone at an average distance from the shoreline (calculated from the crest of the barrier) that is approximatively 2 m whereas in Case 2, the same system of barriers is positioned at the beginning of the surf-zone, at an average distance from the shoreline that is approximatively 4 m.

3.2.1. Case 1

Figure 6a shows the bathymetry in presence of the two submerged breakwaters for the Case 1. A three-dimensional view of the bottom of the curvilinear computational grid is shown in Figure 6b. Figure 6c–d shows two significant cross-sections: one in the gap between the barriers ($y_A = 10.5$ m) and another over the barrier ($y_B = 7.5$ m). The incident wave conditions are the same as the previous test: wave period $T = 1.25$ s and wave height $H = 0.07$ cm.

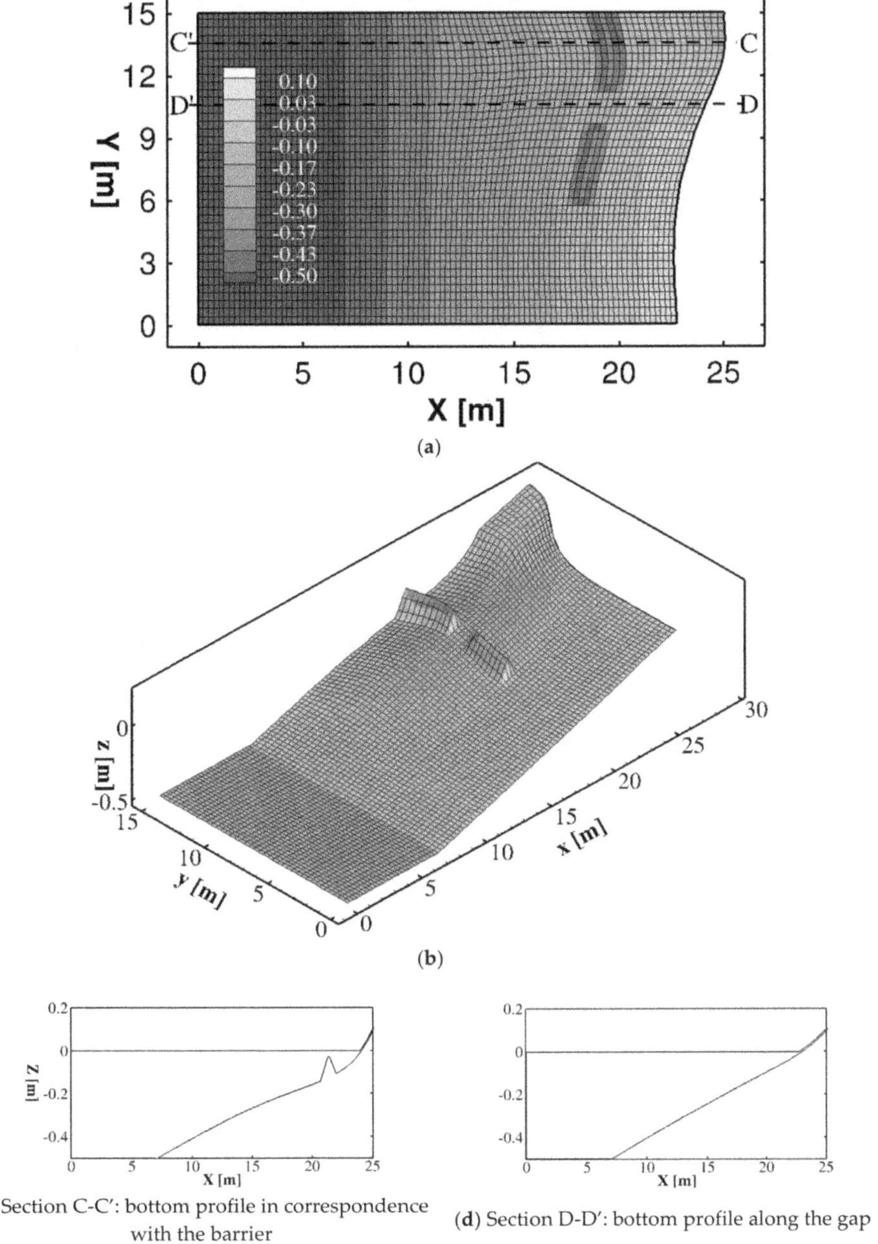

(a)

(b)

(c) Section C-C': bottom profile in correspondence with the barrier

(d) Section D-D': bottom profile along the gap

Figure 6. Case 1: (**a**): Plan view of the curvilinear computational grid (Only one out of every five coordinate lines is shown). (**b**): three-dimensional view of the bottom. (**c–d**): bottom profiles in section C-C' and D-D'.

Figure 7 shows a three-dimensional detail of an instantaneous wave field, in which the nearshore currents are fully developed. The figure shows that the presence of the barriers partially influences the coastal hydrodynamics: the waves that break over the barriers undergo a small reduction in the wave height with respect to those that propagate on the plane beach.

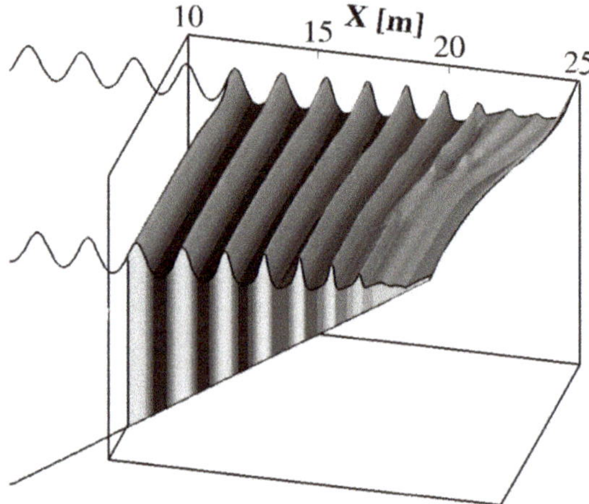

Figure 7. Case 1: Three-dimensional view detail of an instantaneous wave field at the time when the breaking induced circulation is fully developed.

This difference in the wave height is shown in Figure 8a, where the wave height evolution along the two above mentioned sections is presented. Figure 8b shows the time-average (over 120 wave periods) of the cross-shore velocity components calculated near the bottom along the section in the gap between the barriers (section D-D'). In this figure, positive values represent offshore directed velocities.

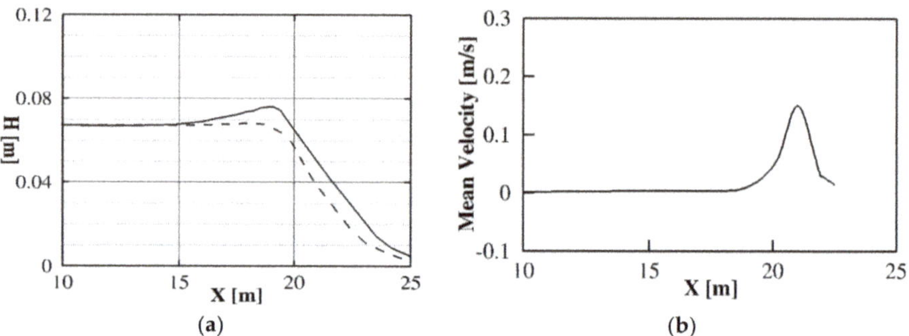

Figure 8. Case 1: (**a**) Wave height comparison between section C-C' (solid line) and section D-D' (dashed line); (**b**) Time-averaged (over 120 wave periods) cross-shore velocity component along the section in the gap between the barriers (section D-D').

Figure 9 shows a plan view of the time-averaged (over 120 wave periods) velocity field near the bottom in which the nearshore currents are fully developed. From this figure it is possible to notice the presence of a circulation pattern characterised by flow velocities that, in the entire lee zone, are directed from the centreline of the barrier to the gaps, where they return offshore. In fact, close to the shoreline, the wave set-up in the gaps is lower than the one in the lee of the barrier; this gradient in the mean water level drives diverging currents that characterise the above mentioned two-cell erosive circulation pattern. In this Case 1, the presence of the barriers does not significantly alter the surf-zone dynamics and consequently is not able to modify the hydrodynamic conditions from erosive to accretive.

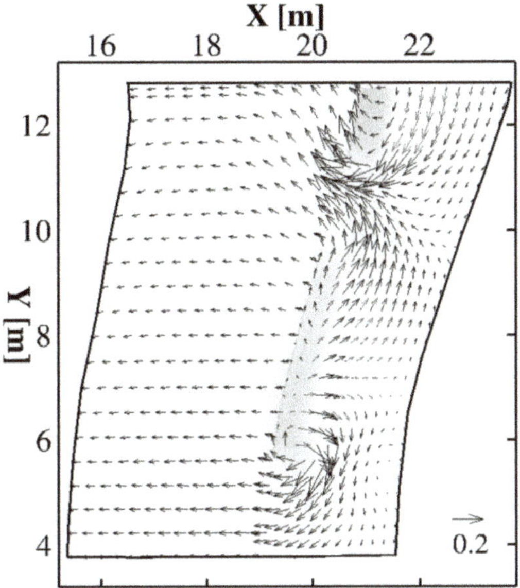

Figure 9. Case 1: Plan view detail of the time averaged (over 120 wave periods) velocity field. Only one out of every four vectors is shown.

3.2.2. Case 2

Figure 10a shows the bathymetry in the presence of the two submerged breakwaters for Case 2. A three-dimensional view of the bottom of the curvilinear computational grid is shown in Figure 10b. Figure 10c–d shows two significant cross sections: one in the gap between the barriers ($y_A = 10.5$ m) and another over the barrier ($y_B = 7.5$ m). The incident wave conditions are the same as the previous case. From Figure 10 it can be noticed that breakwaters are approximatively 2 m more offshore than Case 1.

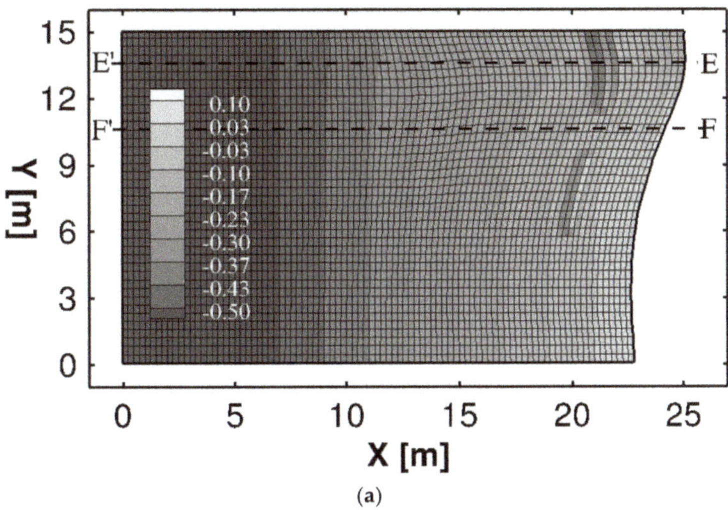

(a)

Figure 10. *Cont.*

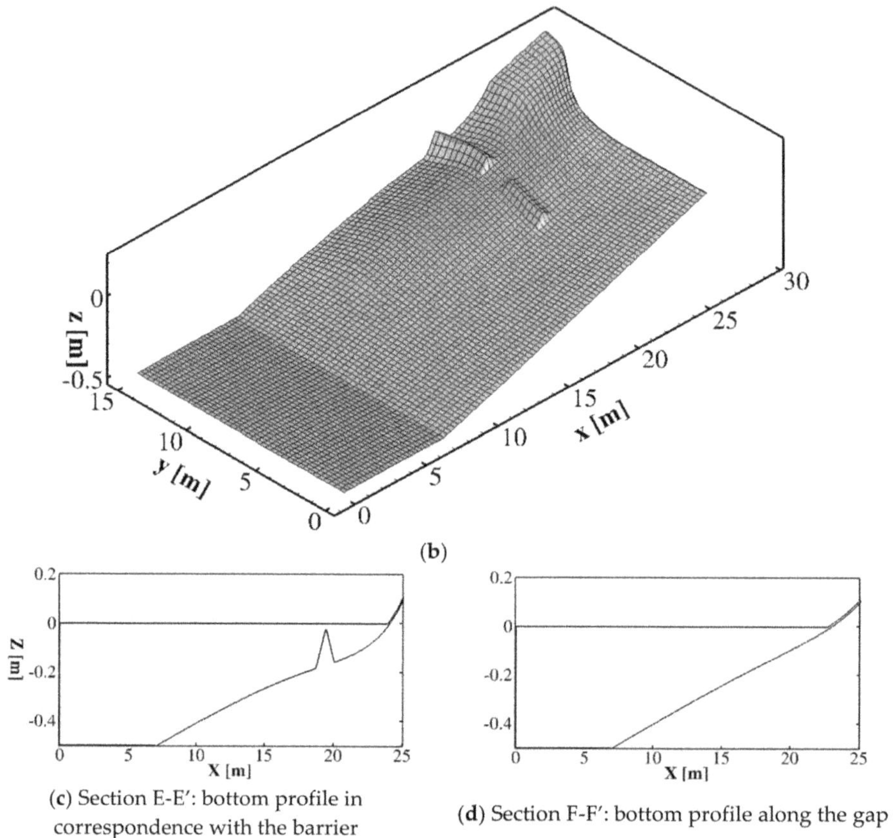

(b)

(c) Section E-E′: bottom profile in correspondence with the barrier

(d) Section F-F′: bottom profile along the gap

Figure 10. Case 2: (**a**): Plan view of the curvilinear computational grid (Only one out of every five coordinate lines is shown). (**b**): three-dimensional view of the bottom. (**c–d**): bottom profiles in section C-C′ and D-D′.

Figure 11 shows a three-dimensional detail of an instantaneous wave field in which the nearshore currents are fully developed. The figure shows that, in this case, the presence of the barriers greatly influences the coastal hydrodynamics: the waves that break over the barriers undergo a significant reduction with respect to those that propagate on the plane beach—the wave that propagates in the gap between the barriers undergoes an increase in the wave height due to the presence of a rip current directed offshore.

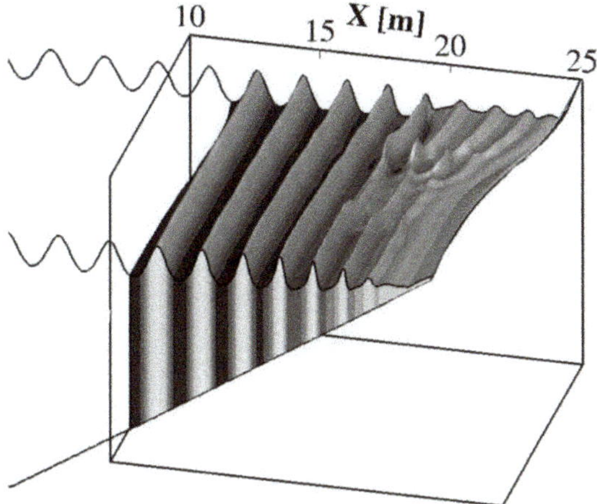

Figure 11. Case 2 Three-dimensional view detail of an instantaneous wave field at the time when the breaking induced circulation is fully developed.

This difference in the wave height is shown in Figure 12a where the wave height evolution along the two above mentioned sections is presented. Figure 12b shows the time-averaged (over 120 wave periods) cross-shore velocity component calculated near the bottom along the section in the gap between the barriers (section F-F′). With respect to Case 1, the significant reduction in the wave height produced by the breakwaters induces a rip current significantly greater than in Case 1.

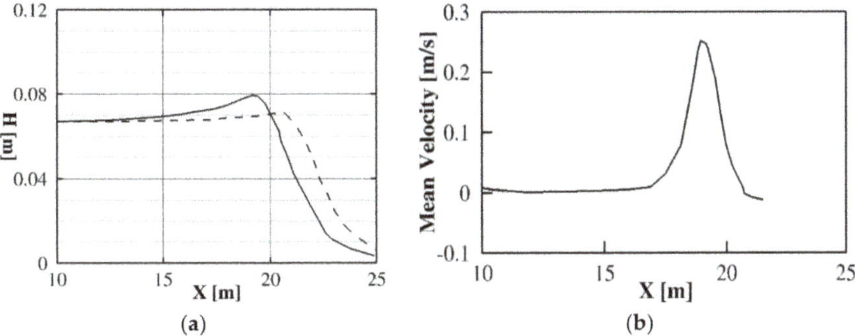

Figure 12. Case 2: (**a**) Wave height comparison between section E-E′ (solid line) and section F-F′ (dashed line); (**b**) Time-averaged cross-shore (over 120 wave periods) velocity component along the section in the gap between the barriers (section F-F′).

Figure 13 shows the vertical time-averaged cross-shore (over 120 wave periods) velocity component respectively at $x = 18.0$ m, $y = 7.5$ m and $x = 20.0$ m, $y = 7.5$ m. As can be observed in Figure 13, the vertical structure of the mean horizontal flow under breaking waves is characterised by onshore directed velocities near the free surface and offshore directed velocities near the bottom (undertow). From these figures, it is possible to deduce that the proposed three-dimensional non-hydrostatic numerical model is able to represent the three-dimensional circulation that occurs, on the offshore side (Figure 13a) and onshore side (Figure 13b) of the submerged breakwater.

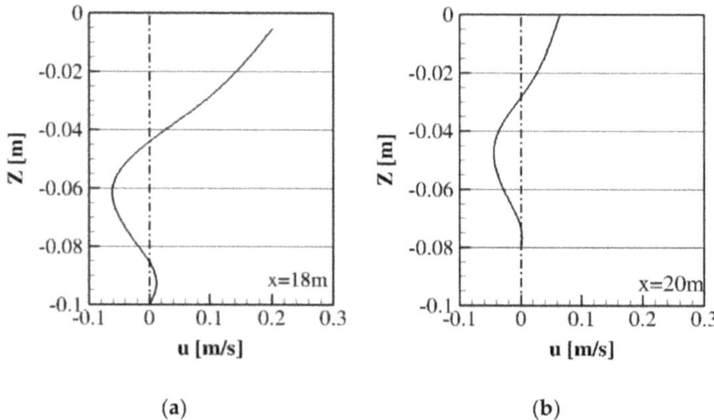

Figure 13. Case 2: Vertical time-averaged (over 120 wave periods) cross-shore velocity profile (a) $x = 18.0$ m, $y = 7.5$ m (b) $x = 20.0$ m, $y = 7.5$ m

Figure 14 shows a plan view of the time-averaged (over 120 wave periods) velocity field near the bottom in which the nearshore currents are fully developed. From this figure, it is possible to notice the presence of a primary circulation characterised by onshore directed flow velocities over the barrier that return offshore at the gaps. In the same figure a secondary circulation it can be seen, opposite the primary one, that takes place near the shoreline. In fact, close to the shoreline, the wave set-up in the gaps is greater than the one in the lee of the barrier—this gradient in the mean water level drives converging currents that characterise the above mentioned four-cell accretive circulation pattern.

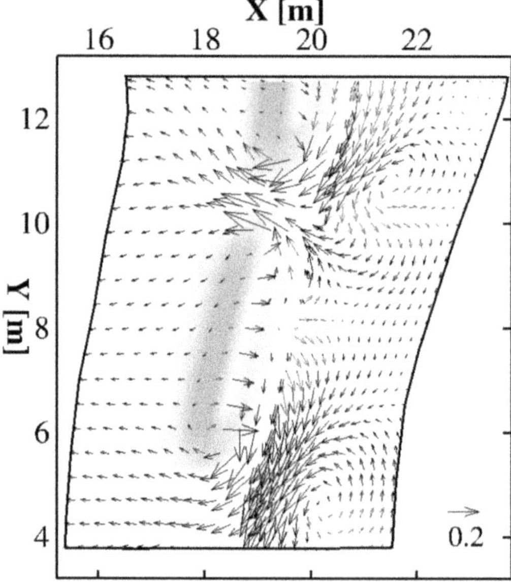

Figure 14. Case 2: Plan view detail of the time averaged (over 120 wave periods) velocity field. Only one out of every four vectors is shown.

From the comparison between Figures 9 and 14 it is possible to deduce that the placement of submerged breakwaters at the beginning of the surf-zone modifies the hydrodynamic conditions from erosive to accretive.

4. Conclusions

A three-dimensional numerical study of the hydrodynamic effects produced by a system of submerged breakwaters in a coastal area with a curvilinear shoreline has been proposed. The three-dimensional numerical model adopted in this paper is based on the integral contravariant formulation of the Navier-Stokes equations in a time-dependent curvilinear coordinate system proposed in [25]. The flow velocity fields and free surface elevation produced by the wave propagation over a system of submerged breakwaters separated by gaps have been simulated. The numerical results show that, for given wave conditions, the positioning of submerged breakwaters in the surf zone induced nearshore currents that can favour the sedimentation of the sediment in the lee of the barriers (accretive circulation pattern). On the contrary, for the same wave conditions, the positioning of the submerged breakwaters too close to the shoreline is not able to modify the circulation pattern from erosive to accretive.

Author Contributions: Conceptualization and methodology F.G. and G.C.; investigation F.P.

Funding: This research received no external funding.

Conflicts of Interest: The authors declare no conflict of interest.

References

1. Van der Meer, J.W.; Briganti, R.; Zanuttigh, B.; Wang, B. Wave transmission and reflection at low-crested structures: Design formulae, oblique wave attack and spectral change. *Coast. Eng.* **2005**, *52*, 915–929. [CrossRef]

2. Williams, H.E.; Briganti, R.; Romano, A.; Dodd, N. Experimental analysis of wave overtopping: A new small scale laboratory dataset for the assessment of uncertainty for smooth sloped and vertical coastal structures. *J. Mar. Sci. Eng.* **2019**, *7*, 217. [CrossRef]

3. Van Doorslaer, K.; Romano, A.; De Rouck, J.; Kortenhaus, A. Impacts on a storm wall caused by non-breaking waves overtopping a smooth dike slope. *Coast. Eng.* **2017**, *120*, 93–111. [CrossRef]

4. Romano, A.; Bellotti, G.; Briganti, R.; Franco, L. Uncertainties in the physical modelling of the wave overtopping over a rubble mound breakwater: The role of the seeding number and of the test duration. *Coast. Eng.* **2015**, *103*, 15–21. [CrossRef]

5. Franco, L.; Geeraerts, J.; Briganti, R.; Willems, M.; Bellotti, G.; De Rouck, J. Prototype measurements and small-scale model tests of wave overtopping at shallow rubble-mound breakwaters: The Ostia-Rome yacht harbour case. *Coast. Eng.* **2009**, *56*, 154–165. [CrossRef]

6. Cannata, G.; Lasaponara, F.; Gallerano, F. Non-linear Shallow Water Equations numerical integration on curvilinear boundary-conforming grids. *WSEAS Trans. Fluid Mech.* **2015**, *10*, 13–25.

7. Cannata, G.; Petrelli, C.; Barsi, L.; Fratello, F.; Gallerano, F. A dam-break flood simulation model in curvilinear coordinates. *WSEAS Trans. Fluid Mech.* **2018**, *13*, 60–70.

8. Lorenzoni, C.; Postacchini, M.; Brocchini, M.; Mancinelli, A. Experimental study of the short-term efficiency of different breakwater configurations on beach protection. *J. Ocean Eng. Mar. Energy* **2016**, *2*, 195–210. [CrossRef]

9. Postacchini, M.; Russo, A.; Carniel, S.; Brocchini, M. Assessing the Hydro-Morphodynamic response of a beach protected by detached, impermeable, submerged breakwaters: A numerical approach. *J. Coast. Res.* **2016**, *32*, 590–602.

10. Hsiao, S.C.; Liu, P.L.-F.; Hwung, H.H.; Woo, S.B. Nonlinear water waves over a three-dimensional porous bottom using Boussinesq-type model. *Coast. Eng. J.* **2005**, *47*, 231–253. [CrossRef]

11. Pelinovsky, E.; Choi, B.H.; Talipova, T.; Woo, S.B.; Kim, D.C. Solitary wave transformation on the underwater step: Asymptotictheory and numerical experiments. *Appl. Math. Comput.* **2010**, *217*, 1704–1718. [CrossRef]

12. Peregrine, D.H. Long waves on a beach. *J. Fluid Mech.* **1967**, *27*, 815–827. [CrossRef]

13. Wei, G.; Kirby, J.T.; Grilli, S.T.; Subramanya, R. A fully nonlinear Boussinesq model for surface waves. Part 1. highly nonlinear unsteady waves. *J. Fluid Mech.* **1995**, *294*, 71–92. [CrossRef]

14. Chen, Q.; Kirby, J.T.; Dalrymple, R.A.; Shi, F.; Thornton, E.B. Boussinesq modeling of longshore currents. *J. Geophys. Res.* **2003**, *108*, 26-1-26-18. [CrossRef]

15. Gallerano, F.; Cannata, G.; De Gaudenzi, O.; Scarpone, S. Modeling bed evolution using weakly coupled phase-resolving wave model and wave-averaged sediment transport model. *Coast. Eng. J.* **2016**, *58*, 1650011-1–1650011-50. [CrossRef]

16. Roeber, V.; Cheung, K.F. Boussinesq-type model for energetic breaking waves in fringing reef environments. *Coast. Eng.* **2012**, *70*, 1–20. [CrossRef]

17. Tatlock, B.; Briganti, R.; Musumeci, R.E.; Brocchini, M. An assessment of the roller approach for wave breaking in a hybrid finite-volume finite-difference Boussinesq-type model for the surf-zone. *Appl. Ocean Res.* **2018**, *73*, 160–178. [CrossRef]

18. Lin, P.; Li, C.W. A σ-coordinate three-dimensional numerical model for surface wave propagation. *Int. J. Numer. Methods Fluids* **2002**, *38*, 1045–1068. [CrossRef]

19. Young, C.C.; Wu, C.H. A σ-coordinate non-hydrostatic model with embedded Boussinesq-type-like equations for modeling deep-water waves. *Int. J. Numer. Methods Fluids* **2010**, *63*, 1448–1470. [CrossRef]

20. Cannata, G.; Petrelli, C.; Barsi, L.; Camilli, F.; Gallerano, F. 3D free surface flow simulations based on the integral form of the equations of motion. *WSEAS Trans. Fluid Mech.* **2017**, *12*, 166–175.

21. Gallerano, F.; Cannata, G.; Lasaponara, F.; Petrelli, C. A new three-dimensional finite-volume non-hydrostatic shock-capturing model for free surface flow. *J. Hydrodyn.* **2017**, *29*, 552–566. [CrossRef]

22. Bradford, S.F. Non-hydrostatic model for surf zone simulation. *J. Waterw. Port Coast. Ocean Eng.* **2011**, *137*, 163–174. [CrossRef]

23. Ma, G.; Shi, F.; Kirby, J.T. Shock-capturing non-hydrostatic model for fully dispersive surface wave processes. *Ocean Model.* **2012**, *43–44*, 22–35. [CrossRef]

24. Cannata, G.; Gallerano, F.; Palleschi, F.; Petrelli, C.; Barsi, L. Three-dimensional numerical simulation of the velocity fields induced by submerged breakwaters. *Int. J. Mech.* **2019**, *13*, 1–14.

25. Cannata, G.; Petrelli, C.; Barsi, L.; Gallerano, F. Numerical integration of the contravariant integral form of the Navier–Stokes equations in time-dependent curvilinear coordinate systems for three-dimensional free surface flows. *Contin. Mech. Thermodyn.* **2019**, *31*, 491–519. [CrossRef]

26. Palleschi, F.; Iele, B.; Gallerano, F. Integral contravariant form of the Navier-Stokes equations. *WSEAS Trans. Fluid Mech.* **2019**, *14*, 101–113.

27. Luo, H.; Bewley, T.R. On the contravariant form of the Navier-Stokes equations in time-dependent curvilinear coordinate systems. *J. Comput. Phys.* **2004**, *199*, 355–375. [CrossRef]

28. Aris, R. *Vectors, Tensors, and the Basic Equations of Fluid Mechanics*; Dover Publications: New York, NY, USA, 1989.

29. Hamm, L. Directional nearshore wave propagation over a rip channel: An experiment. In Proceedings of the 23rd International Conference of Coastal Engineering, Venice, Italy, 4–9 October 1992; pp. 226–239.

30. Ranasinghe, R.; Turner, I.L. Shoreline response to submerged structures: A review. *Coast. Eng.* **2006**, *53*, 65–79. [CrossRef]

31. Ranasinghe, R.; Larson, M.; Savioli, J. Shoreline response to a single shore-parallel submerged breakwater. *Coast. Eng.* **2010**, *57*, 1006–1017. [CrossRef]

32. Haller, M.C.; Dalrymple, R.A.; Svendsen, I.A. Experimental modeling of a rip current system. In Proceedings of the 1997 3rd International Symposium on Ocean Wave Measurement and Analysis, WAVES, Virginia Beach, VA, USA, 3–7 November 1997; pp. 750–764.

33. Haller, M.C.; Dalrymple, R.A.; Svendsen, Ib.A. Experimental study of nearshore dynamics on a barred beach with rip channels. *J. Geophysical Res.* **2002**, *107*, 14-1–14-21. [CrossRef]

Journal of
*Marine Science
and Engineering*

Article

Experimental Observations of Turbulent Events in the Surfzone

Francesca De Serio [1,2,*] and Michele Mossa [1,2]

[1]   Polytechnic University of Bari, DICATECh, 70125 Bari, Italy
[2]   CoNISMa, Inter University Consortium for Marine Sciences, 00196 Rome, Italy; michele.mossa@poliba.it
*   Correspondence: francesca.deserio@poliba.it; Tel.: +39-080-5963557

Received: 6 September 2019; Accepted: 23 September 2019; Published: 24 September 2019

Abstract: In coastal dynamics, large-scale eddies transport and spread smaller turbulent vortices both towards the sea surface, thus contributing to the processes of air-water gas transfer, and towards the sea bottom, inducing sediment pick-up and resuspension. The mechanical role of the breaking-induced vortices to the redistribution of turbulence and turbulent kinetic energy is still unclear and needs a more thorough study, possibly supported by more measurements in this field. Based on this, the present paper aims to investigate the effects of experimental breaking waves in the surf zone. Two regular breaking waves, a spiller and a plunger, which propagate on a fixed slope, were generated in a laboratory channel and were examined shoreward to the breaker line. The measurements of their velocities in the cross-shore plane were assessed by means of a 2D Laser Doppler Anemometer. At the same time and location, elevation data were also acquired using a resistive wave gauge. Here, the principal characteristics are addressed in terms of turbulent intensities, turbulent kinetic energy, length scales and coherent motions. Our results could thus contribute to better define conceptual models used in typical engineering applications in coastal areas.

Keywords: surf zone; plunging breaker; spilling breaker; turbulence intensity; coherent events

---

1. Introduction

Vorticity and turbulence associated with breaking waves make the surf zone of special significance within near-shore dynamics and costal sediment transport [1–6]. Despite the long-term investigations, they are still far from being completely elucidated. Many studies showed that the turbulent flow due to wave breaking is characterized by the formation of large-scale vortex structures which swirl around each other and feature a variety of shapes, named coherent structures [1,7–10]. They primarily control sediment movements and cross-shore morphodynamic evolution. Therefore, predicting near shore morphodynamics requires a thorough knowledge of their formation and spreading throughout the water column, as well as their interactions with bottom sediments. The identification of coherent structures, based on vorticity or other turbulent features, is difficult along with the description of intermittent events of turbulent nature. Many methods have been used, such as quadrant analysis, wavelets, and ensemble averaging to discriminate turbulent fluctuations from the ordered wave motion. However, a thorough knowledge of the formation and spreading of these coherent structures in the surf zone is expected to better explain their interactions with bottom sediments and their effects on a tracers' diffusion [7].

Previous studies of the mechanism controlling the formation and evolution of coherent structures under breaking waves have mainly focused on spilling waves in analogy with bores and jumps, where they were first documented [11,12]. Only a few experimental studies examined this behavior under plunging waves [13,14], where stronger turbulence was observed due to the presence of the overturning jet and the splash-up cycle. Depending on the type of breaker, different types of large-scale coherent vortices can be found in breaking waves [10]. Under a turbulent bore propagating as a spilling wave, Nadaoka et al. [15] first observed large dominant horizontal eddies present in the bore front, while behind the wave crest the flow structure changed rapidly into obliquely downward stretched (i.e., descending) eddies which are responsible for mass flux, and enhancing momentum transport [12]. In a laboratory study of spilling regular waves, Kubo and Sunamura [16] observed, along with obliquely descending vortices, a new type of large-scale turbulence which they called downburst. It is an aerated water mass descending without a great deal of rotation [7], diverging at the bed and producing more sediment movement than the obliquely descending vortices. The counter rotating vortices were also noted in experiments on spilling breaking by Ting and Nelson [17].

Using a large eddy simulation, Christensen and Deigaard [18] and Watanabe et al. [19] have both predicted the formation of counter-rotating vortices extending obliquely downward in plunging and weakly plunging regular waves. Ting and Reinitz [20] observed in experiments on plunging waves the existence of a vortex loop with counter-rotating vorticity generated by the breaking-wave. LeClaire and Ting [21] experienced, in a laboratory, that that large eddies impinging on the bottom was the primary mechanism that lifted the sediment particles into suspension in the plunging case. Further, Ting and Beck [22] with their experiments, and Zou et al. [23] with numerical runs, found that for a plunging wave, the dominant sediment suspension mechanism was the outward and upward deflected flow field which originated by the splashing jet impinging at the bottom.

Ting [7] observed that in a solitary breaking wave, a downburst of turbulence descends and diverges at the bed creating two counter-rotating vortices. More recently, Lubin et al. [24] numerically simulated plunging breaking waves and their results confirmed what shown in a rare documentary footage where breaking waves were recorded from underwater. In fact, they computed 3D vortical tubes, like vortex filaments elongated in the main flow direction, occurring under the plunging breaker, connecting the splash-up and the main tube of air. Lubin et al. [25] recently observed them also experimentally. Furthermore, Mukaro [26] observed that instantaneous vorticity tended to organize into thin filaments of counter-rotating pairs.

Therefore, it is evident that further experimental investigation on coherent events is still needed, in the attempt to better clarify these mechanisms of turbulence spreading due to wave breaking. This is the motivation of the present study, which investigates the surf zone of two laboratory regular waves, a plunging one and a spilling one, propagating on a fixed and impermeable slope in a 2D wave flume. The measurements were carried out by means of a Laser Doppler Anemometer (LDA) in some selected channel sections, located in the outer and inner surf zone. At the same time and locations, the elevation data were also acquired using a resistive probe. It is worth noting that, compared with previous research described above, the present work differs in particular for the measurement instrumentation adopted and the analyzed target region. Moreover, here the principal characteristics are discussed in terms of turbulent kinetic energy and turbulent intensities, length scales and coherent motions, for both spilling and plunging breaking, with the aim of contributing to a better understanding of turbulence dispersion. In this way, the conceptual models that describe typical engineering applications in coastal areas could benefit from this research.

2. Materials and Methods

2.1. Experimental Set Up

The present experiment was carried out in the wave flume of the Department of Civil, Environmental, Land, Building Engineering and Chemistry (DICATECh) of the Polytechnic University of Bari (Italy). The flume was 45 m long, 1 m wide and 1.20 m deep, provided with glass sidewalls and bottom (Figure 1). A fixed and uniform slope 1:20 was created in the channel, starting 12 m from the paddles. The slope was 33 m long. The still water level in the channel was $h_0 = 0.7$ m above the horizontal bed. The examined regular waves had heights H at the paddle respectively equal to 10 cm and 7 cm. Breaking occurred at a depth of about 11 cm for both tests. The plunging wave case, named

Test_P, is characterized by a Irribarren number ξ_0 equal to 0.6, while the spilling wave case, named Test_S, by $\xi_0 = 0.2$ (see [2] for details).

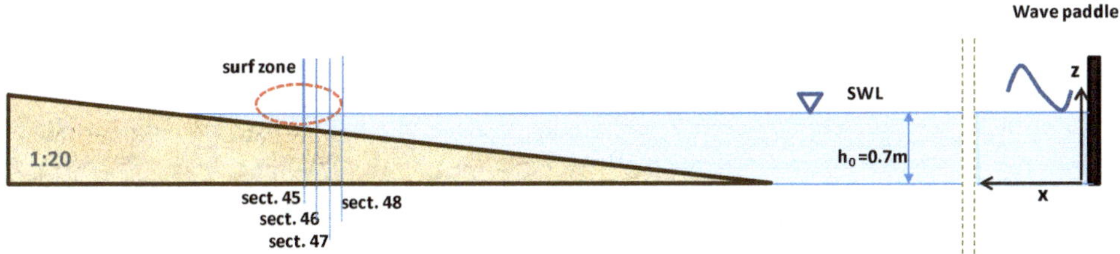

Figure 1. Sketch of the wave flume, with the location of the examined vertical sections.

The measurements were carried out in some selected channel sections in the shoaling and in the breaking zone, for both tests. In the present study, four vertical sections close to the breaking region (identified as section 48, section 47, section 46 and section 45, in Figure 1) are examined. Specifically, section 48 was in the prebreaking region, section 47 was where the incipient breaking occurred, while in sections 46 and 45, the wave re-arranged into a bore. The horizontal reference system x has origin at the wave paddle and is positive towards the shore, while the vertical axis z has origin on the bottom and is positive upwards. The locations and local depths h for sections 48, 47, 46 and 45 are displayed in Table 1. In the same table, the wave height H, the wave period T, the length computed according to Airy theory L_A, and the Ursell parameter, an indicator of the wave non-linear behavior, are also shown.

Table 1. Main characteristics of the two tested regular wave trains.

TEST	Sect.	x [m]	h [cm]	H [cm]	T [s]	L_A [m]	Ur	Measurements Range [cm]
Test_P plunging	48	22.9	14	12.64	4	4.66	1000	$z = 1$–11
	47	23.3	11.3	9.67	4	4.19	1177	$z = 1.3$–8.3
	46	23.7	10	7.02	4	3.72	970	$z = 2$–8
	45	24.1	8.5	5.98	4	3.17	976	$z = 1.5$–7
Test_S spilling	48	22.9	14	6.78	1	1.06	27.88	$z = 1$–12
	47	23.3	11.3	6.62	1	0.97	43.58	$z = 3.3$–9.3
	46	23.7	10	5.52	1	0.89	43.74	$z = 1$–8
	45	24.1	8.5	4.21	1	0.78	41.94	$z = 1$–7.5

The instantaneous horizontal u and vertical w velocities, respectively measured along the longitudinal x and the vertical z axis, were assessed by means of a 2D back-scatter, four-beam LDA along the water column. These velocities are conventionally established positive if respectively directed onshore and upwards. The acquisition frequency of the LDA was mainly approximately 100 Hz and the duration of each measurement was at least 200 s, thus permitting sufficient wave cycles (on average 50) to use in the subsequent phase-averaging procedure. At the same time, a resistive wave gauge aligned with the LDA volume of measurement was used to measure the surface elevation. In each section, the velocities were assessed vertically in points spaced 0.5 cm or 1 cm, depending on local conditions (Table 1).

The used LDA instrumentation could provide a single point measurement, differently from the more sophisticated instrumentation, i.e., PIV [1,17] and Volumetric 3V [20,22], which instead provided measurements respectively in 2D slices and 3D volumes, thus allowing a direct description of a large portion of the flow field. In any case, the data processing allowed the indirect detection of the main turbulent structures, as observed in previous studies. Moreover, it is worth noting that the sampling frequency of the adopted LDA system is much greater than that of the PIV and V3V systems utilized in previous research (ranging from 7.5 Hz to 15 Hz). This aspect is fundamental because it guarantees to better capture the turbulent behavior of the phenomenon at higher frequencies.

2.2. Operational Procedure

Several techniques can be used to extract eddies in a flow field [7,27], including: direct analysis of the vorticity field as computed from velocity map; analysis of the velocity gradient tensor which has one imaginary eigenvalue whose distribution is associated with regions of local swirling motion; LES filtering, based on low pass filtering of velocity data to remove small scale contributions; wavelet transform algorithms of the velocity vector field; quadrant analysis.

The present study first used the phase-averaging procedure (i.e., the most common notation for periodic signal processing) to extract the turbulent velocity components from the instantaneous velocity time series [3,28]. As well known, the instantaneous signal can be regarded as the sum of a time-averaged part (addressed with the capital letter), a phase-averaged or orbital part (addressed with the subscript *ph*) and a turbulent fluctuation (addressed with the prime sign). As an example, for the horizontal velocity, being t the variable time, it can be written as:

$$u(t) = U + u_{ph} \, (t/T) + u'(t) \tag{1}$$

After this, the focus is on the quadrant analysis to identify the coherent motions and quantify the bursting processes in the instantaneous turbulent velocity time series, based on [29]. The decomposition of the turbulent Reynolds shear stress $u'w'$ (apart from $-\rho$, water density) is considered into four quadrants defined by the Cartesian axes of the scatter plot $w' = f(u')$. The quadrants identify four types of events: (Q1) outward interactions, $u' > 0$ and $w' > 0$; (Q2) ejections, $u' < 0$ and $w' > 0$; (Q3) inward interactions, $u' < 0$ and $w' < 0$; (Q4) sweeps, with $u' > 0$ and $w' < 0$. Associated with the bursting process are ejections, where low-speed fluid penetrates the high flow region, and sweeps, where the high-speed fluid penetrates the low flow region [30]. Both ejections and sweeps contribute positively to the Reynolds stress. An event is included in the consideration if the product $u'w'$ is larger than a critical threshold, otherwise it is discarded as noise. Setting this threshold is somewhat arbitrary.

Therefore, based on previous research [29], the coherent event has been defined as the one responding to the following condition: $|u'w'| \geq (\mu + \sigma)$, where μ is the mean and σ the standard deviation of the $u'w'$ time series measured in the investigated point. Analogously, an intense coherent event is detected when $|u'w'| \geq (\mu + 3\sigma)$.

3. Results and Discussion

3.1. Phase-Averaged Distributions

To provide an insight in the flow field, Figure 2 sketches the phase-averaged velocities (u_{ph}, w_{ph}) in a wave cycle, at the depths examined in Test_P (Figure 2a) and Test_S (Figure 2b) in section 47, i.e., where the breaking is incipient. The scale of the velocities is the same in both Figure 2a,b. Since no noticeable deformation of the wave shape is observed through propagation over a distance of one wavelength, at least offshore the measuring section, the horizontal axis in Figure 2 (showing the instant time in the period) may be regarded approximately as the horizontal spatial coordinate in the wavelength [7,15].

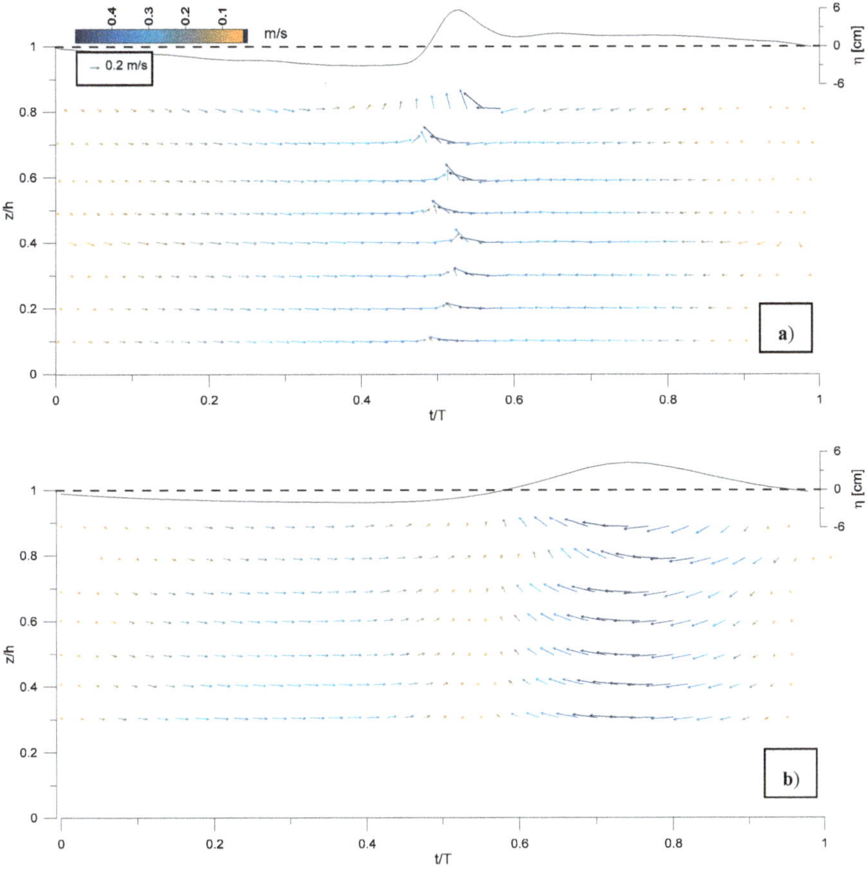

Figure 2. Section 47: phase-averaged velocities (u_{ph}, w_{ph}) in the wave cycle at all investigated depths for (**a**) Test_P and (**b**) Test_S.

In the upper part of each plot is also added the phase-averaged elevation η. A difference is immediately noticed in the vertical asymmetry of the incipient breaking wave between the two tests. As expected, in Test_P the wave crest is higher while the wave trough is lower than in Test_S. Moreover, the horizontal asymmetry highlights that in Test_P, the steep plunging breaker occupies a much more limited portion of the wavelength, compared to the case of the spilling breaker. Nevertheless, the distribution along the wave cycle of the phase-averaged velocities, plotted in Figure 2 as vectors with components (u_{ph}, w_{ph}), is quite similar for Test_P and Test_S. Indeed, it resembles the wave elevation trend in both tests, with increasing velocities in the transition trough-crest (ascending phase) and decreasing velocities in the transition crest-trough (descending phase), at all depths. As well, a rotation of the velocity vectors is observed in the wave cycle at all depths, for both tests, coherently with the curvature of the surface, as already noted in previous experiments [7,31]. Consequently, during the first descending phase, the velocities are fairly uniformly oriented downwards and offshore, while in the accelerating phase, they rotate upward and onshore. In the plunging wave (Figure 2a) below the wave crest, the maximum velocity value is noted, which is quite vertical in the most superficial layer. In the spilling case (Figure 2b), below the crest, onshore horizontal velocity is observed at all depths. The reduction in the velocity values (absolute values) from the surface to the bottom is more marked in Test_P (Figure 2a) than in Test_S (Figure 2b).

Referring to the same section 47, it is worth depicting the distribution in the wave cycle of the turbulence intensities u'_{ph} and w'_{ph}, computed by definition as the square root of $(u'^2)_{ph}$ and $(w'^2)_{ph}$. Figure 3a provides the turbulence intensities for Test_P, showing that the horizontal ones are higher than the vertical ones, as well as their decrease from surface to bottom, consistently with the incipient breaking. They both have their peak at the lower edge of the eddy region, also in accordance with Nadaoka et al. [15]. Moreover, the tail of the contour lines is slightly oriented downwards and offshore, thus endorsing the possible presence of obliquely descending eddies (ODEs) numerically and experimentally proved by Lubin [10,24].

Figure 3b displays u'_{ph} and w'_{ph} values for Test_S, highlighting that in the spilling case they are approximately one third of those measured in Test_P. Furthermore, in this case, the u'_{ph} values are greater than the w'_{ph} values and they both reduce towards the bottom. The highest values for u'_{ph} are localized under the ascending and descending phases of the wave, while the highest values for w'_{ph} are localized under the crest. These patterns seem consistent with the propagating spilling bore. It is worth noting the recursive occurrence for both the tests of the turbulence intensities along the wave phases, thus stating an organized turbulent structure, which is quite typical for plunging and spilling waves.

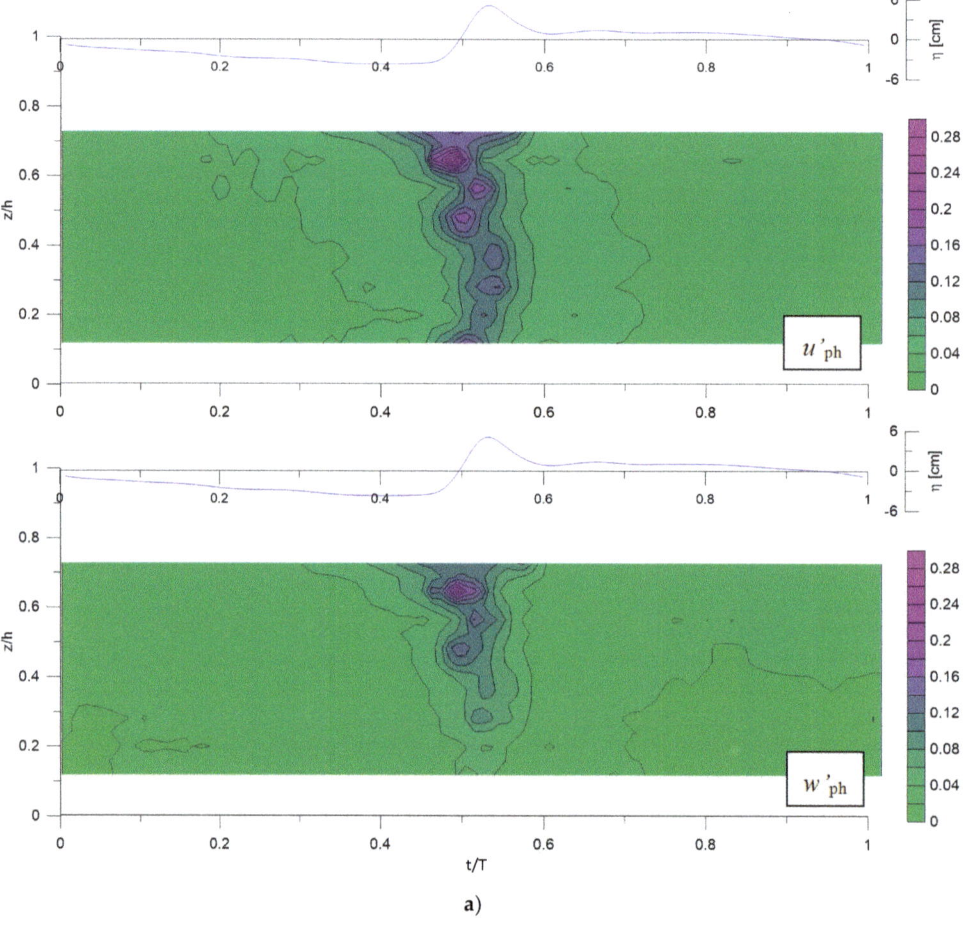

a)

Figure 3. *Cont.*

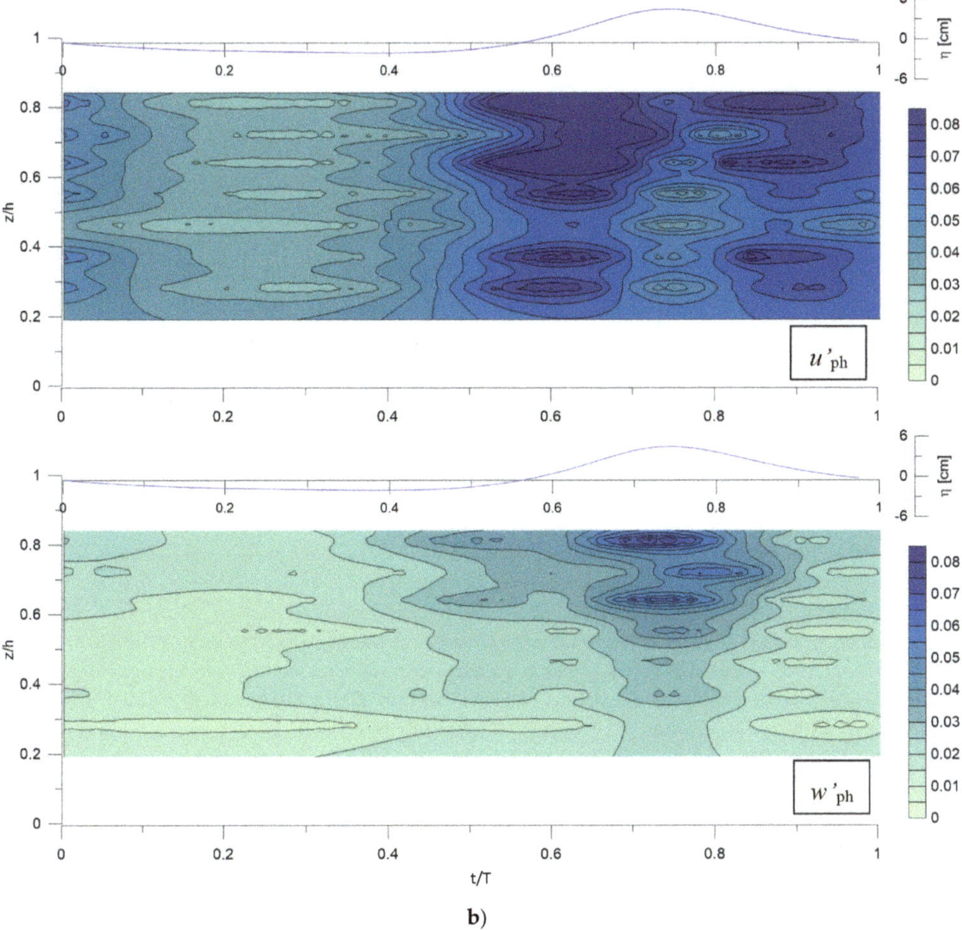

b)

Figure 3. Section 47: phase-averaged turbulent intensities in the wave cycle at all the investigated depths: (**a**) u'_{ph} and w'_{ph} for Test_P; (**b**) u'_{ph} and w'_{ph} for Test_S. Units are in m/s.

3.2. Time-Averaged Profiles

The vertical profiles of the time-averaged turbulent Reynolds shear stresses $U'W'$ and of the turbulent kinetic energy K are examined in sections 48, 47, 46 and 45. For Test_P, they are respectively shown in Figure 4a,b, where the plotted points refer to the locations below the still water level.

The turbulent Reynolds stresses $U'W'$ in Figure 4a are quite negligible for sections 48 and 47, excepting the lowest point of section 48, in the boundary layer streaming, where the $U'W'$ value is positive, coherently with the effect of the bottom boundary layer. Referring in particular to section 47, the small order of magnitude of the $U'W'$ stresses (Figure 4a) can be compared with the higher one of the u'_{ph} and w'_{ph} shown in Figure 3a. It can be deduced that the major contribution to the total turbulent Reynolds stress in the incipient breaking (section 47) is essentially due to the phase-averaged turbulent velocities.

In section 46, immediately after the breaking, the $U'W'$ stresses are negative near the surface, and almost zero in the underlying region. In section 45 where the wave is reforming, they assume negative values near the surface and the highest positive values in the intermediate and lower part. This vertical gradient of $U'W'$ suggests a downward transport of turbulence from the surface, due to the occurred plunger [3,28], which is more evident in the upper region, for $z/h > 0.4$ in both sections 46 and 45. The largest positive $U'W'$ stresses are present in section 45, when the wave has formed again, near the bottom, proving that the effects of the bottom boundary layer are again felt [3,28,31].

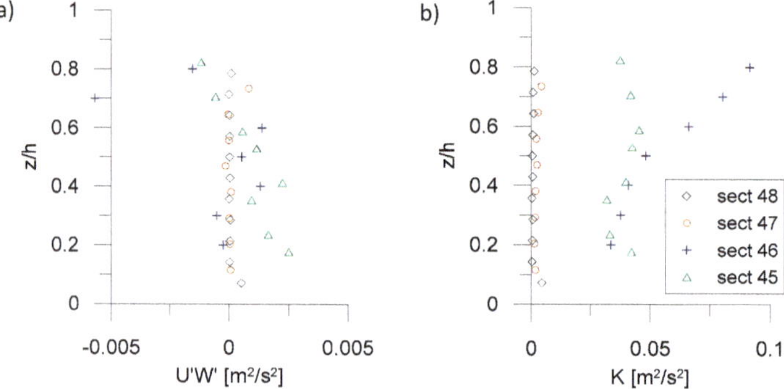

Figure 4. Test_P: vertical profiles of the time-averaged turbulent Reynolds stresses (**a**) and turbulent kinetic energy (**b**).

The different behavior between the prebreaking/incipient breaking sections (sections 48 and 47) and the sections shoreward the breaking (sections 46 and 45) is also evident in the vertical trends of the time-averaged turbulent kinetic energy K, here computed following Svendsen [32]. In fact, it tends to zero in sections 48 and 47 (Figure 4b), while its highest values are assessed in section 46, especially for $z/h > 0.5$ due to the reversed plunger, where a decreasing trend towards the bottom is noted, meaning a spreading downward as in [1]. Analogously, also in section 45, the K profile has a similar shape, but the surface values are lower than in section 46, considering that the turbulence is decreasing with the increasing distance from the plunging breaker, meaning spreading downstream, in accordance with [1].

These observations are consistent for Test_P with the vertical distributions of the time-averaged integral length scales, Lx and Lz, associated with the turbulent large eddies (Figure 5a,b) respectively in x and z directions. The integral and turbulent length scales have been computed by multiplying the local time-averaged velocity respectively for the turbulent time scale and the integral time scale. In turn, both turbulent and the integral time scales have been estimated based on the temporal autocorrelation function of the turbulent velocity fluctuations (for the detailed formulation, please refer to [33]). It is noteworthy that the integral length scales Lx, while having order O(0.1 h) in both sections 48 and 47 and 46, tend to be of order O(0.2 ÷ 0.3 h) in section. 45, thus proving that large eddies are stretched in the x direction, as a consequence of the breaking.

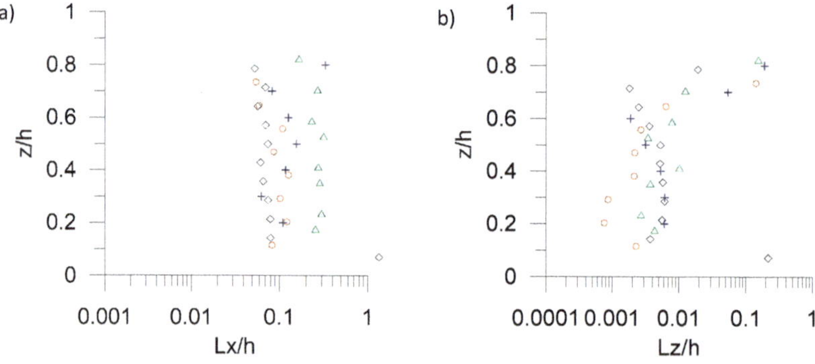

Figure 5. *Cont.*

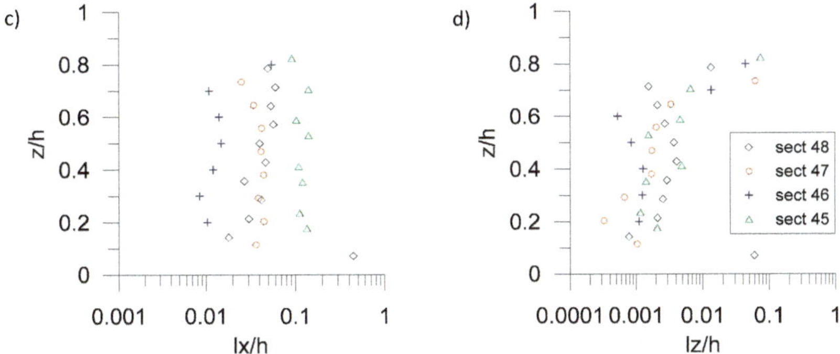

Figure 5. Test P: time-averaged horizontal (**a**) and vertical (**b**) integral length scales; time-averaged horizontal (**c**) and vertical (**d**) turbulent length scales.

Moreover, by comparing Figure 5a,c, it can be deduced that in sect. 46, immediately after the breaking, the *lx* scales are one order lower than the *Lx* scales. The same is also noted referring to the vertical *Lz* and *lz* scales (Figure 5b,d), thus confirming the transfer of the energy contained at large scales to small scales, where it is turned into heat by viscosity [7]. Moreover, for $z/h > 0.6$ both *Lz* and *lz* scales show increasing values as they approach the surface, with order ranging from O(0.01 *h*) up to O(0.1 *h*), consistently with the formation of the overturning plunging jet.

The same analysis of the time-averaged quantities has been performed also for Test_S. Analogously to Test_P, the *U'W'* stresses are quite negligible in sections 48 and 47 (Figure 6a). In section 46, when the spilling wave has just broken, there is a uniform positive trend along the *z*-axis. Onshore, in section 45, the *U'W'* profile shows the same bi-triangular shape already noted in Test_P (Figure 4a), although in this case, lower positive values are displayed near the bottom and higher negative values for $z/h > 0.3$.

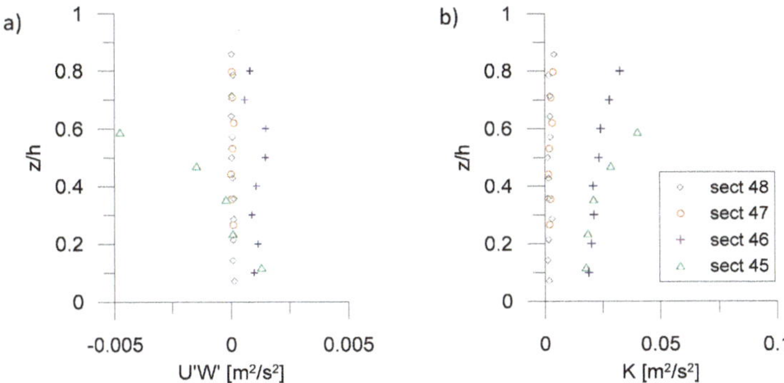

Figure 6. Test_S: vertical profiles of the time-averaged turbulent Reynolds stresses (**a**) and turbulent kinetic energy (**b**).

Furthermore, the distribution of *K* (Figure 6b) shows vertical trends already observed in Test_P, with decreasing values from the surface towards the bed in both sections 46 and 45. Nevertheless, in Test_S, the values of *K* are halved with respect to Test_P. Moreover, the higher values of *K* are observed in section 45, based on the presence of the bore after spilling, rather than in section 46. Comparing Figure 6a,b with Figure 4a,b, it can be assumed that with the bore propagating onshore, the spilling breaker continues to feed turbulence in the surface region, unlike the plunging wave, which produces higher turbulence but is located in a more compact area, where the breaking occurs. Moreover, the turbulent kinetic energy is less intense in the spilling case than in the plunging case.

Referring to the time-averaged integral length scales Lx and Lz (Figure 7a,b), for each section there is a difference of approximately two orders of magnitude between them, proving the presence of eddies much more elongated in the horizontal direction than in the vertical direction. This horizontal stretching is thus greater for this spilling case than for the plunging one. Referring to the turbulent length scales lx and lz for Test_S, although they are lower than the corresponding integral ones, lx has a comparable order of magnitude with Lx and the same thing happens when comparing lz and Lz. Consequently, their graphs are not shown for brevity, being of little significance especially in the semi-logarithmic version. In any case, it is worth noting that despite eddies could not be observed directly, as for example in [1,17,20,22] based on PIV or V3V acquisitions, the distributions of these length scales permitted the indirect deduction of the presence of eddies in the target domain.

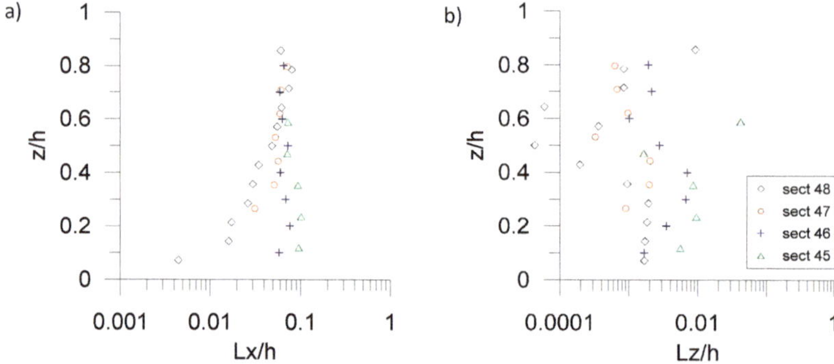

Figure 7. Test_S: time-averaged horizontal (**a**) and vertical (**b**) integral length scales.

3.3. Quadrant Analysis

The results of the quadrant analysis are here described only referring to section 47, for the sake of brevity, for both Test_P (Figure 8) and Test_S (Figure 9). Figures 8 and 9 display the quadrant plots of the instantaneous turbulent velocities (u', w') at five different phase intervals. The phases for which the data are plotted are written for each plot and have been selected based on meaningful parts along the wave cycle. Namely, for Test_P (Figure 8): $0 < t/T < 0.4$ descending phase; $0.4 < t/T < 0.6$ descending phase up to wave trough; $0.6 < t/T < 0.65$ ascending phase up to wave crest; $0.65 < t/T < 0.75$ steep descending phase; $0.75 < t/T < 1$ flat descending phase. Analogously, for Test_S (Figure 9): $0 < t/T < 0.2$ descending phase; $0.2 < t/T < 0.4$ flat descending up to wave trough; $0.4 < t/T < 0.6$ gradual ascending phase; $0.6 < t/T < 0.8$ ascending phase up to wave crest; $0.8 < t/T < 1$ descending wave back. The phase-averaged wave elevation is also inserted in both figures to facilitate the identification of the abovementioned phase intervals. Further, the results in the top row refer to a near surface point, while those in the bottom row to a near bottom point. The scaling for the axes is the same for an easier comparison, except the interval 0.6–0.75 in Test_P (Figure 8) due to high turbulent velocities.

The top row of Figure 8 shows that in section 47 for the point at $z/h = 0.7$, the turbulent horizontal fluctuations u' are the largest during the ascending phase $0.6 < t/T < 0.65$ (confirming [29]) and under the wave crest $t/T = 0.65$. On the contrary, they are minimal in the descending phase ($0.65 < t/T < 0.75$), where the turbulent vertical fluctuations w' prevail and are the largest. In the remaining wave cycle, i.e., for $0 < t/T < 0.6$ and $0.75 < t/T < 1$, the plots tend to assume a quite circular shape, meaning that the ($u'w'$) points are fairly evenly distributed in all the quadrants Q1 ÷ Q4. For $0.4 < t/T < 0.6$, the cloud of points is slightly elongated along the horizontal axis, thus the u' components prevail. For $0.75 < t/T < 1$ the cloud of points is slightly aligned along Q2 and Q4 quadrants, proving the presence of ejections and sweep after the wave crest passage.

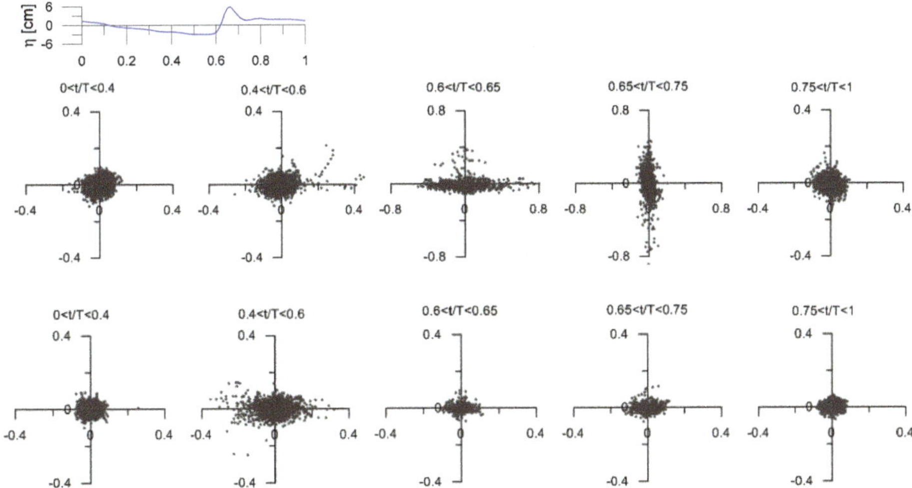

Figure 8. Test_P, sect. 47: quadrant plots in five phase intervals, at z/h = 0.7 (**top row**) and z/h = 0.2 (**bottom row**).

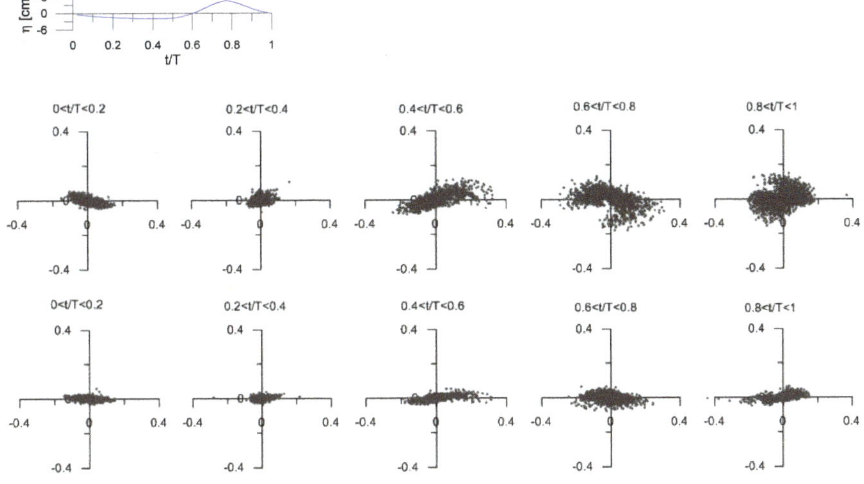

Figure 9. Test_S, section 47: quadrant plots in five phase intervals, at z/h = 0.7 (**top row**) and z/h = 0.3 (**bottom row**).

The bottom row of Figure 8, referring to the point at z/h = 0.2 in section 47, depicts generally flattened plots. This is consistent with the presence of the bottom, reducing w' even if a large eddy due to wave breaking reaches the bottom boundary, as also noted in [29]. The u' in $0.6 < t/T < 0.75$ are much smaller than the corresponding ones near the surface, meaning that the passage of the wave crest has a negligible effect near the bottom. On the contrary, increasing values of u' near the bottom are noted rather than near the surface, during the descending phase and below the wave trough ($0.4 < t/T < 0.6$). From this analysis, a recursive tendency of the bursts in the wave cycle is evident, especially in the superficial layer, referring to the trough-crest transition and to the descending phase, under the back face of the wave.

The top row of Figure 9 displays the quadrant analysis assessed for the spilling case (Test_S) in section 47, referring to the point at z/h = 0.7. In the descending phase $0 < t/T < 0.2$, the plot is sloped along Q2–Q4 quadrant, proving the presence of both ejections and sweeps. In the region approaching the trough ($0.2 < t/T < 0.4$), turbulent intensities reduce strongly, while they increase again during the ascending phase ($0.4 < t/T < 0.6$) where both outward and inward motions are noted (cloud deployed along Q1–Q3 quadrant). During the ascending phase up to the wave crest ($0.6 < t/T < 0.8$), the plot is directed along Q2–Q4, with the presence of ejections and sweeps, already noted in Figure 3b. In the back face of the wave ($0.8 < t/T < 1$), again both the inward and outward motions are relevant (Q1–Q3).

Similar to the bottom row of Figure 8, also in the bottom row of Figure 9, referring to the point at $z/h = 0.3$, there are flattened clouds, which indicate the effect played by the bottom on the w' velocities. Furthermore, the largest w' values during the interval $0.6 < t/T < 0.8$ are noted, when the crest is approaching. The largest u' values are observed in the transition trough-crest-trough ($0.4 < t/T < 1$). In the case of the spilling wave, the quadrant analysis highlights a dependence of the bursts on the wave phases even more marked with respect to the plunging case. Although different from the plunging case, in the spilling case, this effect is strongly noted close to the surface as well as close to the bottom.

Finally, the authors detected the coherent events, according to the method described in Section 3.2. For both Test_P and Test_S, in each examined section, for three points (near surface, intermediate, near bottom), the assessed time series of $u'w'$ have been analyzed. The mean and the standard deviation were computed for each considered $u'w'$ signal and then the number of cases for which $|u'w'| \geq (\mu + \sigma)$ were detected, thus identifying the coherent events in the series. The same has been made for detecting the most intense coherent events, for the condition $|u'w'| \geq (\mu + 3\sigma)$ [24].

Table 2 summarizes the results for the most significant sections, i.e., sections 47, 46 and 45. In the last two columns, the authors indicate: the percentage of coherent events %CE i.e., the number of data for which $|u'w'| \geq (\mu + \sigma)$, rated by the total number of the acquired data, as well, the percentage of intense coherent events %IE, i.e., the number of data for which $|u'w'| \geq (\mu + 3\sigma)$, rated by the total number of the acquired data.

Table 2. Coherent and intense coherent events detected for TEST_P and TEST_S by means of the threshold values for the signal $u'w'$.

	TEST_P				TEST_S		
Sect.	z/h	%CE	%IE	Sect.	z/h	%CE	%IE
47	0.7	3.2	1.1	47	0.8	13.6	2.6
47	0.5	2.9	1.1	47	0.4	13.3	2.2
47	0.2	7.4	1.9	47	0.3	12.5	2.2
46	0.7	10.5	1.6	46	0.7	22.5	1.4
46	0.5	7.5	1.2	46	0.5	24.7	1.6
46	0.2	5.5	1.1	46	0.3	21.8	1.6
45	0.7	12.3	1.8	45	0.5	18.6	2.3
45	0.5	13.7	1.7	45	0.2	20.0	1.6
45	0.2	11.9	1.7	-	-	-	-

For Test_P, the following can be noted. In section 47, the percentage of both coherent and intense coherent events increases from the surface towards the bottom. On the contrary, immediately after breaking in section 46, both percentages of coherent and intense coherent events, decrease with depth.

In section 45, the %CE percentage is the highest at the intermediate depth, while the %IE percentage remains quite unchanged at all depths. The comparison between sections 46 and 45 leads to thoughts of a spreading of the coherent events towards the bottom, while the intense ones are more localized in the upper layer, which is consistent with the splashed jet.

For Test_S, in section 47 the percentage of both coherent and intense coherent events decreased from the surface towards the bottom, being opposite to what was observed in Test_P in the same section, and presumably due to the formation of the roller. In section 46, the maximum %CE percentage is at mid-depth, while the %IE percentage tends to increase with depth. In section 45, the %CE percentage increases towards the bottom, while the %IE percentage decreases towards the bottom. It could be argued from the comparison of percentages in section 47 and 46 that in this case, the coherent and intense coherent events, initially more localized around the spiller, are spread downwards and onshore, thus affecting also section 45. The percentages computed are comparable with previous results [23,30].

Finally, it is observed that in Test_P, coherent and intense coherent events are more infrequent, especially referring to sections 46 and 45, where the percentages for Test_S are 2–3 times higher than in Test_P. This could be explained by the different breaking mechanism and the different nature of generated turbulence that characterize the two waves, especially considering the non-deterministic nature of a splashing jet.

4. Conclusions

Two regular breaking waves, a spiller and a plunger, reproduced in a laboratory wave flume have been examined. Specifically, this study focused on the turbulent structure induced by their breaking in the surf zone. By separating time-averaged velocities, phase-averaged velocities and turbulent velocities, as well as by adopting the quadrant analysis, turbulent eddies and coherent bursts were detected. The principal results of the study are summarized below.

The distribution along the wave cycle of the phase-averaged velocities at breaking is similar for both plunging and spilling cases, considering that their trends resemble the wave elevation, despite the nonlinear behavior in the surf zone, with increasing velocities in the transition trough-crest and decreasing velocities in the transition crest-trough at all depths. The rotation of the velocity vectors in the wave cycle at all depths is also noted coherently with the curvature of the surface and in agreement with previous experiments [7,31]. The pattern of the phase-averaged turbulence intensities at the incipient breaking highlights for both plunging and spilling cases that the horizontal ones are greater with respect to the vertical ones, as well as their decrease from surface to bottom, as expected. Rather, for the plunging case, these turbulence intensities spread downwards and offshore, thus endorsing the possible presence of obliquely descending eddies proved in previous studies [10,15,24].

For both plunging and spilling waves, the vertical profiles of the time-averaged turbulent Reynolds shear stresses show that they are fairly negligible in the prebreaking and incipient breaking region, while they assume relevance in sections immediately after breaking, showing a vertical gradient that suggests a downward transport of turbulence from the surface. The vertical trends of the time-averaged K display that turbulence is higher in the plunger than in the spiller, but it decreases with increasing depth as well as with increasing distance from the breaking section. Differently, the bore propagating onshore the spilling breaker seems to feed turbulence in the surface region also onshore on the breaking section. The different order of magnitudes of the time-averaged integral and turbulent length scales onshore breaking confirms that the energy contained at large scales is transferred to small scales. Furthermore, these values (breaking onshore) prove the presence of eddies much more elongated in the horizontal direction than in the vertical one. This horizontal stretching is greater for the spilling case than for the plunging case.

Finally, the quadrant analysis has been executed to detect coherent events, recurring in the wave cycle. For both the plunging and the spilling test, a reduction of the intensities of the bursts is evident from the surface towards the bottom. In the plunging wave, maxima horizontal u' values are noted in the ascending phase of the wave, and maxima vertical w' values in the descending phase, when the breaking is incipient. In the spilling case, there are greater turbulent bursts corresponding to ejections and sweeps mainly during the ascending phase and below the crest of the wave. Different from the plunging case, in the spilling the dependence of the bursts on the wave phases is strongly noted close to the surface as well as close to the bottom. By examining the number of coherent and intense coherent events based on a previous study [29], it can be argued that, due to the different breaking mechanism, the spilling case is dominated by coherent events while in the plunging case, the splashing jet has a much more non-deterministic nature, thus inducing less frequent coherent events.

Author Contributions: Conceptualization and methodology F.D.S. and M.M.; data analysis and writing, F.D.S.; review and editing, F.D.S. and M.M.

Funding: This research received no external funding.

Acknowledgments: The technical staff of the laboratory at DICATECh is gratefully acknowledged.

Conflicts of Interest: The authors declare no conflicts of interest.

References

1. Kimmoun, O.; Branger, H. A particle images velocimetry investigation on laboratory surf-zone breaking waves over a sloping beach. *J. Fluid Mech.* **2007**, *588*, 353–397. [CrossRef]
2. De Serio, F.; Mossa, M. Experimental study on the hydrodynamics of regular breaking waves. *Coast. Eng.* **2006**, *53*, 99–113. [CrossRef]
3. De Serio, F.; Mossa, M. A laboratory study of irregular shoaling waves. *Exp. Fluids* **2013**, *54*, 1536. [CrossRef]
4. De Padova, D.; Brocchini, M.; Buriani, F.; Corvaro, S.; De Serio, F.; Mossa, M.; Sibilla, S. Experimental and Numerical Investigation of Pre-Breaking and Breaking Vorticity within a Plunging Breaker. *Water* **2018**, *10*, 387. [CrossRef]
5. Van der Zanden, J.; van der A, D.A.; Cáceres, I.; Hurther, D.; McLelland, S.J.; Ribberink, J.S.; O'Donoghue, T. Near-Bed Turbulent Kinetic Energy Budget Under a Large-Scale Plunging Breaking Wave Over a Fixed Bar. *J. Geophys. Res. Oceans* **2018**, *123*, 1429–1456. [CrossRef]
6. Van der A, D.A.; van der Zanden, J.; O'Donoghue, T.; Hurther, D.; Cáceres, I.; McLelland, S.J.; Ribberink, J.S. Large-scale laboratory study of breaking wave hydrodynamics over a fixed bar. *J. Geophys. Res. Oceans* **2017**, *122*, 3287–3310. [CrossRef]
7. Ting, F.C.K. Large-scale turbulence under a solitary wave: Part 2. Forms and evolution of coherent structures. *Coast. Eng.* **2008**, *55*, 522–5368. [CrossRef]
8. Zhou, Z.; Sangermano, J.; Hsu, T.J.; Ting, F.C.K. A numerical investigation of wave-breaking-induced turbulent coherent structure under a solitary wave. *J. Geophys. Res. Oceans* **2014**, *119*, 6952–6973. [CrossRef]
9. Farahani, R.J.; Dalrymple, R.A. Three-dimensional reversed horseshoe vortex structures under broken solitary waves. *Coast. Eng.* **2014**, *91*, 261–279. [CrossRef]
10. Lubin, P.; Glockner, S. Numerical simulations of three-dimensional plunging breaking waves: Generation and evolution of aerated vortex filaments. *J. Fluid Mech.* **2015**, *767*, 364–393. [CrossRef]
11. Longo, S. Vorticity and intermittency within the pre-breaking region of spilling breakers. *Coast. Eng.* **2009**, *56*, 285–296. [CrossRef]
12. Huang, Z.C.; Hwung, H.H.; Chang, K.A. Wavelet-based vortical structure detection and length scale estimate for laboratory spilling waves. *Coast. Eng.* **2010**, *57*, 795–811. [CrossRef]
13. Lim, H.J.; Chang, K.A.; Huang, Z.C.; Na, B. Experimental study on plunging breaking waves in deep water. *J. Geophys. Res. Oceans* **2015**, *120*, 2007–2049. [CrossRef]
14. Na, N.; Chang, K.A.; Huang, Z.C.; Lim, H.J. Turbulent flow field and air entrainment in laboratory plunging breaking waves. *J. Geophys. Res. Oceans* **2016**, *121*, 1–30. [CrossRef]
15. Nadaoka, K.; Hino, M.; Koyano, Y. Structure of the turbulent flow field under breaking waves in the surf zone. *J. Fluid Mech.* **1989**, *204*, 359–387. [CrossRef]
16. Kubo, H.; Sunamura, T. Large-scale turbulence to facilitate sediment motion under spilling breakers. In Proceedings of the IV Conference on Coastal Dynamics, Lund, Sweden, 11–15 June 2001; pp. 212–221.
17. Ting, F.C.K.; Nelson, J.R. Laboratory measurements of large-scale near-bed turbulent flow structures under spilling regular waves. *Coast. Eng.* **2011**, *58*, 151–172. [CrossRef]
18. Christensen, E.D.; Deigaard, R. Large eddy simulation of breaking waves. *Coast. Eng.* **2001**, *42*, 53–86. [CrossRef]
19. Watanabe, Y.; Saeki, H.; Hosking, R.J. Three-dimensional vortex structures under breaking waves. *J. Fluid Mech.* **2005**, *545*, 291–328. [CrossRef]
20. Ting, F.C.K.; Reimnitz, J. Volumetric velocity measurements of turbulent coherent structures induced by plunging regular waves. *Coast. Eng.* **2015**, *104*, 93–112. [CrossRef]

21. LeClaire, P.D.; Ting, F.C.K. Measurements of suspended sediment transport and turbulent coherent structures induced by breaking waves using two-phase volumetric three-component velocimetry. *Coast. Eng.* **2017**, *121*, 56–76. [CrossRef]

22. Ting, F.C.K.; Beck, D.A. Observation of sediment suspension by breaking-wave-generated vortices using volumetric three-component velocimetry. *Coast. Eng.* **2019**, *151*, 97–120. [CrossRef]

23. Zhou, Z.; Hsu, T.J.; Cox, D.; Liu, X. Large-eddy simulation of wave-breaking induced turbulent coherent structures and suspended sediment transport on a barred beach. *J. Geophys. Res. Oceans* **2017**, *122*, 207–235. [CrossRef]

24. Lubin, P.; Glockner, S.; Kimmoun, O.; Branger, H. Numerical study of the hydrodynamics of regular waves breaking over a sloping beach. *Eur. J. Mech. B/Fluids* **2011**, *30*, 552–564. [CrossRef]

25. Lubin, P.; Kimmoun, O.; Véron, F.; Glockner, S. Discussion on instabilities in breaking waves: Vortices, air-entrainment and droplet generation. *Eur. J. Mech. B/Fluids* **2018**, *73*, 144–156. [CrossRef]

26. Mukaro, R. Vorticity filaments beneath regular turbulent flow. *J. S. Afr. Inst. Civil. Eng.* **2017**, *59*, 2–10. [CrossRef]

27. Camussi, R. Coherent structure identification from wavelet analysis of particle image velocimetry data. *Exp. Fluids* **2002**, *32*, 76–87. [CrossRef]

28. Van der Werf, J.; Ribberink, J.; Kranenburg, W.; Neessen, K.; Boers, M. Contributions to the wave-mean momentum balance in the surf zone. *Coast. Eng.* **2017**, *121*, 212–220. [CrossRef]

29. Truong, S.H.; Uijttewaal, W.S.J.; Stive, M.J.F. Exchange Processes Induced by Large Horizontal Coherent Structures in Floodplain Vegetated Channels. *Water Resour. Res.* **2019**, *55*, 2014–2032. [CrossRef]

30. Cox, D.T.; Kobayashi, S. Identification of intense, intermittent coherent motions under shoaling and breaking waves. *J. Geophys. Res.* **2000**, *105*, 14223–14236. [CrossRef]

31. Melville, W.K.; Veron, F.; White, C.J. The velocity field under breaking waves: Coherent structures and turbulence. *J. Fluid. Mech.* **2002**, *454*, 203–233. [CrossRef]

32. Svendsen, I.A. Analysis of surf zone turbulence. *J. Geophys. Res.* **1987**, *92*, 5115–5124. [CrossRef]

33. Nezu, I.; Nakagawa, H. *Turbulence in Open-Channel Flows*; Balkema/CRC Press: Boca Raton, FL, USA, 1993.

 Journal of
Marine Science and Engineering

Article

Combining Numerical Simulations and Normalized Scalar Product Strategy: A New Tool for Predicting Beach Inundation

Matteo Postacchini [1,*] and Giovanni Ludeno [2]

[1] Department of Civil and Building Engineering and Architecture, Università Politecnica delle Marche, 60131 Ancona, Italy
[2] Institute for Electromagnetic Sensing of the Environment, National Research Council, I-80124 Napoli, Italy; ludeno.g@irea.cnr.it
* Correspondence: m.postacchini@staff.univpm.it

Received: 10 July 2019; Accepted: 17 September 2019; Published: 19 September 2019

Abstract: The skills of the Normalized Scalar Product (NSP) strategy, commonly used to estimate the wave field, as well as bathymetry and sea-surface current, from X-band radar images, are investigated with the aim to better understand coastal inundation during extreme events. Numerical simulations performed using a Nonlinear Shallow-Water Equations (NSWE) solver are run over a real-world barred beach (baseline tests). Both bathymetry and wave fields, induced by reproducing specific storm conditions, are estimated in the offshore portion of the domain exploiting the capabilities of the NSP approach. Such estimates are then used as input conditions for additional NSWE simulations aimed at propagating waves up to the coast (flood simulations). Two different wave spectra, which mimic the actual storm conditions occurring along the coast of Senigallia (Adriatic Sea, central Italy), have been simulated. The beach inundations obtained from baseline and flood tests related to both storm conditions are compared. The results confirm that good predictions can be obtained using the combined NSP–NSWE approach. Such findings demonstrate that for practical purposes, the combined use of an X-band radar and NSWE simulations provides suitable beach-inundation predictions and may represent a useful tool for public authorities dealing with the coastal environment, e.g., for hazard mapping or warning purposes.

Keywords: Nonlinear Shallow-Water Equations; numerical simulation; marine radar; Normalized Scalar Product; wave-field estimate; bathymetry estimate; beach inundation

---

1. Introduction

The nearshore area hosts many human activities and interests, which have become increasingly important in recent years. Climate change can have a potentially destabilizing effect on such activities: both sea level rise and increase of sea storminess may promote an increase of the coastal inundation frequency, this severely affecting engineering works, as well as ecological, recreational and environmental issues [1,2]. Hence, analyses of coastal vulnerability and inundation risk are required worldwide to prevent and/or mitigate such changes, and represent a fundamental task to be accounted for when coastal plans and policies are carried out [3–7].

Specifically, coastal flooding and erosion can potentially induce important economic losses and human fatalities. An Integrated Coastal Zone Management (ICZM) approach, which spans over the main aspects of the coastal region (from prediction to protection, from engineering to ecosystems, from tourism to sustainability) seems essential to mitigate the mentioned negative impacts. The need for an integrated approach in the coastal area is globally acknowledged, also in view of the increasing pressures which are featuring our world, e.g., population growth, concentration of economic activities, expanding tourism [8]. As an example, an important protocol has been signed in the framework of the Barcelona Convention, which promotes an integrated approach of the coastal zone management in the Mediterranean area [9]. Such protocol focuses on the spatial development of coastal zones, and the need for assessing and measuring pressures from human activities. Studies have been undertaken based on the proposed approach and applications have been attempted, e.g., [10,11].

In this framework, estimating the beach inundation of a specific location is fundamental for, e.g., hazard mapping purposes, coastal risk analysis, application of the ICZM approach. Suitable tools are thus required for the monitoring of the nearshore region, with many devices being tested for this purpose in the recent years. For instance, Lagrangian drifters and floaters have long been used for oceanic studies and can properly measure hydrodynamics (waves, currents) and seabed morphology along the coast, e.g., [12,13]. Furthermore, fixed instruments (Eulerian approach) deployed in the sea for a certain time are often used for the reconstruction of the wave field in the nearshore region, with the possibility to record data during both calm and stormy periods, though the risk concerning the equipment safety, e.g., [14,15]. However, devices based on a remote sensing approach, such as acoustic sensors [16], X-band radar [17], video-monitoring systems [18,19], look easier to be used during severe conditions. Diaz et al. [20] assessed a reliable measure when two or more of such devices are contemporary used and their results are combined. Suitable post-processing tools are also required to retrieve useful information in the analyzed domain, e.g., wave characteristics (height, period, direction), breaker features (roller length, energy dissipation), seabed morphology.

Among the remote sensing systems mentioned above, we have here considered the X-band marine radar, which provides an alternative to the standard detection systems of the sea state (e.g., buoys and others). Currently, the X-band marine radar, besides being used for the detection and tracking of the targets on sea surface, is employed as a remote sensing tool for sea-state monitoring both on ships [21–23] and in coastal areas [24–27]. The radar echoes arising from the sea surface and received by antenna radar, if properly processed, allow one to retrieve the sea-state parameters such as length, period and direction of the dominant waves, and the significant wave heights, as well as to reconstruct the sea-surface current and bathymetry field.

With a specific focus on the bathymetry reconstruction, various radar imaging techniques are available, and these significantly evolved in time. In the last three decades, several approaches for current and bathymetry estimation have been developed as Least-Square (LS) [28], Iterative Least-Square (ILS) [29], Normalized Scalar Product (NSP) [30,31]. Since many of the above tools, such as X-band radars, are not able at measuring or reconstructing the hydro-morphodynamics in the shallowest region of the beach, numerical modeling can be used as an additional tool to be exploited for coastal inundation purposes. Specifically, the results/estimates obtained using the remote sensor can be used as input conditions for the numerical model, i.e., a "chain approach", which is similar to that applied in recent studies. These demonstrate the feasibility to reconstruct the shallow-water hydrodynamics induced by forcing actions of various nature exploiting a series of numerical/analytical models and/or sea observations [7,32,33].

Typical numerical approaches for the description of the coastal regions are based on the solution of either the Nonlinear Shallow-Water Equations (NSWE) [34,35] or the Boussinesq equations [36,37]. On one hand, both model/equation types are based on the assumption of wave nonlinearity, a fundamental feature which describes the wave evolution in intermediate to shallow waters. On the other hand, the wave frequency dispersion, which mainly describes the wave energy re-distribution during the wave propagation, also plays an important role in the description of the wave transformation in the nearshore region, but is only accounted for by Boussinesq-type models. However, shallow-water computations provide suitable results even when a NSWE solver is used, often exploiting a reduced computational cost. Recently, with the aim to exploit the best skills from both approaches, based on the use of both Boussinesq (in intermediate to deep waters) and NSWE (in the shallowest waters) approaches, hybrid models have been built [38,39].

The present paper investigates the potentialities of the combination of the NSP approach and shallow-water simulations in properly predicting the beach inundation. In detail, the NSP estimates are used as boundary conditions (wave-field reconstruction) and initial conditions (bathymetry reconstruction) of a Nonlinear Shallow-Water Equations (NSWE) solver, able at transferring shoreward the reconstructed wave and providing the beach inundation and runup. Different wave conditions and various beach scenarios have been tested, with the purpose to illustrate and validate a novel method to evaluate the coastal inundation exploiting radar images.

The paper is divided as follows. Section 2 describes the used tools, i.e., the hydrodynamic NSWE solver and the NSP approach, as well as their combination. The results obtained using the above tools are reported and discussed in Section 3. A final discussion closes the paper (Section 4).

2. Materials and Methods

The present work is aimed at illustrating a methodology that can be applied for the prediction of the beach inundation. Specifically, remote sensing devices, aimed at reconstructing the bathymetric evolution of a coastal region [20,28–31], may also be exploited to reconstruct the wave field in the investigated domain, so as to generate suitable initial and boundary conditions to be used by phase-resolving models which propagate the wave up to the swash zone. With such purpose, a NSWE solver is first used to generate the wave field over a real-world bathymetry to simulate what actually occurs in the open sea. The generated wave field is then modulated to simulate the signal recorded by an X-band marine radar. An NSP algorithm is used to reconstruct both wave field and bathymetry. Such reconstructions are then used to run numerical simulations of beach inundation.

2.1. The hydrodynamic Solver

The numerical model used for the wave-field generation within the domain (baseline simulations), as well as for the wave propagation from the boundary condition to the shore (flood simulations), is that described in [34], which is based on the solution of the NSWE, i.e., depth-averaged equations of the mass and momentum conservation. The non-conservative form of the NSWE, which is based on the Cartesian coordinate system (x, y, z), follows:

$$d_{,t} + (ud)_{,x} + (vd)_{,y} = 0, \tag{1}$$

$$u_{,t} + uu_{,x} + vu_{,y} + gd_{,x} = gh_{,x} - B_x, \tag{2}$$

$$v_{,t} + uv_{,x} + vv_{,y} + gd_{,y} = gh_{,y} - B_y, \tag{3}$$

with $\mathbf{v} = (u, v)$ the depth-averaged horizontal velocity, g the gravity acceleration, d the total water depth, h the still water depth, both providing the surface water level $\eta = d - h$. The friction contributions are provided through a Chezy-type approach, i.e., $B_x = C_f|\mathbf{v}|u/d$ and $B_y = C_f|\mathbf{v}|v/d$, where the dimensionless coefficient C_f is described using a constant value (as in many previous works [7,34,35]).

Due to the depth-averaged nature of the NSWE model, the inversion technique exploits the shallow-water approximation of the dispersion relation, i.e.,

$$\omega = k\sqrt{gh} + \bar{k} \times \overline{U}, \tag{4}$$

where ω is the angular frequency, $\bar{k} = (k_x, k_y)$ the wave vector, $k = |\bar{k}|$ the wave number and $\overline{U} = (U, V)$ the external current. However, since (4) leads to nonunique solutions when the inversion technique is applied to estimate both depth and current, the present simulations do not account for external currents ($|\overline{U}| = 0$ m/s) and the depth is the only unknown. Hence, the only currents in the numerical domain are those inherently generated by the wave field (for more details, see [31]).

2.2. The NSP Method

The step between baseline and flood simulations is undertaken using the NSP approach, which has already been tested and validated using both real-world and numerical data [25,31,40], and is typically used to elaborate radar images. These are not direct representation of the wave elevation profile, but they are tied to the slopes of the long sea waves (tilt modulation), to the roughness of the riding short waves (hydrodynamic modulation). Moreover, the radar signals are affected by the "shadowing", for which no information can be recovered for the sea spots that are not in the line of sight (LOS) [17,41–43]. These phenomena are called modulation effects and are introduced during the microwave remote sensing process. Herein, only the shadowing and tilt modulation effects are considered. In particular, the shadowing effect can be interpreted as a geometrical phenomenon and is modeled as if the radar antenna was not actually receiving any signal from the shadowed parts of the sea surface. Instead, the tilt modulation depends on the local power received by the radar antenna on the slope of the observed surface. This causes a modulation of the received radar signal, which is dependent on the angle between the radar illumination ray (i.e., the local LOS) and the normal vector to the wave surface. More details on the geometry and formulation of the tilt model and the shadowing can be found in [17,31]. As in [31], the numerical analysis assumes an incoherent marine radar placed at a height of 20 m above the mean sea level and at coordinates $(x, y) = (0, 0)$ m.

The strategy employed to retrieve both sea-state parameters and bathymetry in coastal areas from the simulated X-band radar images, involves a spatial partitioning of the radar images into partially overlapping sub-areas [21,24,44,45]. Therefore, each radar image belonging to the temporal sequence under analysis is partitioned into N_s spatially overlapping sub-areas, giving rise to N_s temporal subsequences. Once the data partitioning step is performed, N_s radar spectra $\{F^j(\bar{k}, \omega)\}_{j=1,...,N_s}$ corresponding to the sub-areas, are computed via the Fast Fourier Transform (FFT) algorithm. Then, starting from the local radar spectra [17,31], the local bathymetric value $\hat{h}^j$ is founded on the maximization of the following Normalized Scalar Product, i.e.,

$$\hat{h}^j = \arg\max_h \frac{\langle |F^j(\bar{k}, \omega)|, G(\bar{k}, \omega, h) \rangle}{\sqrt{P_{F^j}}} \tag{5}$$

where $G(\bar{k}, \omega, h) = \delta(\omega - k\sqrt{gh})$ is the characteristic function (being $\delta(\cdot)$ the Dirac-delta distribution) based on the dispersion relation, $\langle |F^j|, G \rangle$ represents the scalar product of the functions $|F^j|$ and G, while P_{F^j} is the power associated with $|F^j|$. It is worth noticing that the NSP strategy allows us the joint estimation of the surface currents and the bathymetry from the radar spectrum [25,46], with the wave fields generated using a null sea-surface current for the present context.

Therefore, the bathymetry field can thus be reconstructed starting from the local estimates. Such information is of course extremely useful for various coastal applications, but it is also an essential tool to correctly estimate the wave field since, as is well known, the depth, as well as the sea-surface current, are required to define a Band-Pass (BP) filter to separate the energy of the sea signal from the noise background in the radar spectrum. The required sea-wave spectrum $F_w^j(\bar{k}, \omega)$ can be obtained from the filtered image spectrum $F_I^j(\bar{k}, \omega)$ by resorting to the radar Modulation Transfer Function (MTF), which mitigates the distortions affecting the radar echoes and caused by the acquisition geometry (e.g., shadowing and tilt modulation) [17,24,25,47]. For this purpose, the MTF described in [17] and reported in the following equation has been adopted.

$$F_w^j(\bar{k}, \omega) = \frac{F_I^j(\bar{k}, \omega)}{k^\beta} \quad \text{for } \beta = 1.2 \tag{6}$$

2.3. The Methodology

The methodology which is here proposed aims at combining the application of remote sensors, such as X-band marine radar, for the reconstruction of both offshore wave field and bathymetry, with the numerical modeling in shallow waters. The main goal is that of providing a useful tool for coastal authorities to either reconstruct the nearshore hydrodynamics or predict the coastal inundation. In this case, the above tool needs to be properly integrated and act in real time such as warning systems. The methodology described in the following can be used to reach this purpose.

- The zero-th step is the wave field to be investigated in a specific region of interest: while this exists in the real world, it is here simulated to provide realistic conditions (baseline simulations), similarly to [31].
- The first step is the reconstruction of both wave field and bathymetry in an offshore portion of the domain (order of kilometers from the coast, where typically radar sequence data are collected) using the NSP approach.
- The estimated bathymetry and wave characteristics (significant wave height H_{m0}, peak period T_p, direction θ_0) are used to build, respectively, the seabed morphology and the boundary conditions (in terms of JONSWAP spectra) at different depths/locations.
- Flood simulations are run using boundary conditions at depths of either $h = 5$ m or $h = 9$ m, and using either ground-truth bathymetries or equilibrium-profile bathymetries. While the use of a ground-truth bathymetry represents the case of beach surveys available for the region of interest, the equilibrium profiles may be applied when surveys are not available, as such an approach has already been observed to provide an accurate description of coastal dynamics [48,49].
- The beach inundation obtained from each baseline simulation is compared to the results of the corresponding flood simulations.

Figure 1 summarizes the main steps of the methodology: what is illustrated in the present work (left) and what is expected to do in the real world (right).

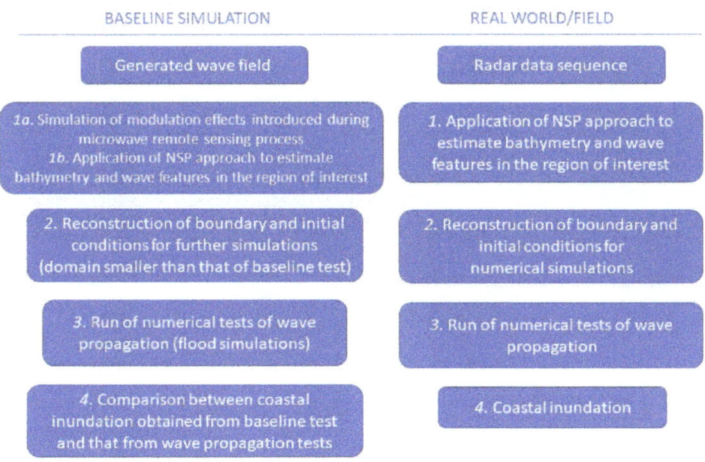

Figure 1. Sketch of the methodology applied in the present work (**left**) or to be applied to the real world (**right**).

2.4. The Application

The described methodology has been applied to one of the most representative coastal areas of the central Italy. Specifically, the coast of Senigallia, one of the most important touristic towns of the Italian Middle Adriatic coast, has been recently studied under several points of view. In particular, the site is characterized by the combination of several forcing actions and processes, especially due to the Misa River (length of ~48 km, watershed of ~383 km^2), whose estuarine area is highly engineered and separates the Northern coast of Senigallia, protected by an array of emerged breakwaters, from the Southern part, a natural open coast [15,50]. Such natural, unprotected beach is part of the longest unprotected beach of the Marche Region, extending for about 12 km, from the Misa River estuary to some kilometers North of the Esino River estuary.

The nearshore region used in the present work is located in the Southern beach, where a couple of video-monitoring systems have been recently installed, with the aim to reconstruct the beach dynamics in a large portion of the Southern beach [51]. The swash-zone slope ranges between 1:30 and 1:40. Moving seaward, a multiple array of submerged bars exists in a water depth $h = (0–3)$ m, while a mild slope ($\sim$ 1:200) characterizes larger depths. Fine to medium sands, i.e., $d_{50} = (0.125–0.5)$ mm, is typical of the emerged beach, while fine sand, i.e., $d_{50} = (0.125–0.25)$ mm, characterizes the submerged beach [50].

Recent studies undertaken on the Southern coast underline that the most frequent and energetic waves are those coming either from ESE or NNE. The former waves are forced by Levante-Scirocco winds, the latter by Bora winds, and both types correspond to the predominant waves of the investigated coastal region. It has also been observed that different wave climates provide different morphological behavior of the submerged beach, mainly concerning the bar migration and evolution. Specifically, the NNE waves experience a reduced refraction and are characterized by a small storm-surge, as a consequence of the reduced fetch due to the elongated shape of the Adriatic basin. Conversely, the refraction process of ESE waves is important, and the longer fetch produces a larger surge. Such considerations translate into a significantly different hydrodynamics, e.g., NNE waves are steeper than ESE waves, and morphodynamics, e.g., larger bar smoothing during NNE waves [50].

The Numerical Simulations

The bathymetry of Senigallia surveyed in May 2013 has been used for the baseline simulations, i.e., the beach developing from the Misa River estuary for about 1 km to the South (see Figure 2). Since such survey was undertaken up to a depth of about 6 m, it was extended seaward, assuming a constant slope of 1:200 up to a 10 m depth.

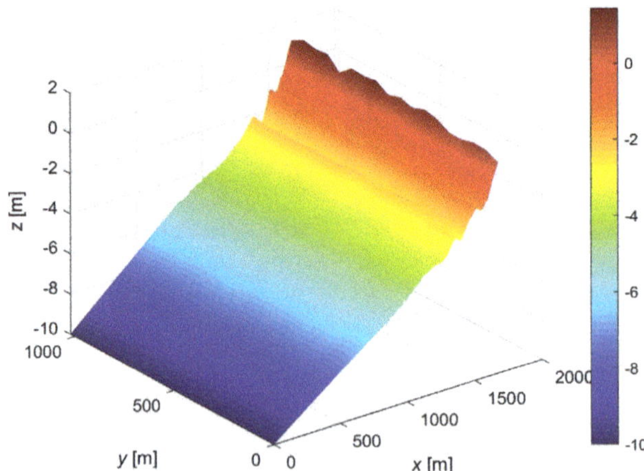

Figure 2. Original ground-truth bathymetry used for the baseline simulations.

The baseline tests have been run using two different types of waves which characterize the energetic features of the considered site, as described in [50]. Specifically, for the application of the described methodology, one simulation reproduces a milder wave coming from ESE, the other a steeper wave coming from NNE. The main wave characteristics (significant height H_{m0}, peak period T_p, wave direction θ_0) used for the generation of the boundary conditions at the offshore side of the numerical domain are shown in Table 1. It is worth noting that the direction is also affected by the refraction process, which enables the wave rays to rotate while approaching the coast. A local angle is also introduced, i.e., the angle at which the waves enter the domain. This is estimated as $\theta_{0,l} = \theta_0 - \theta_{coast}$, where $\theta_{coast} \approx 45°$ is the angle between the normal to the coast and the North. As with the numerical simulations presented in [31] and in agreement with recent field observations of swell waves along the Senigallia coast during calm states and storm tales [52], long-crested waves with no directional spreading and $T_p = (5–10)$ s have been reproduced.

Table 1. Wave characteristics used at the offshore boundary of the numerical domain for the baseline simulations.

Wave Type	$H_{m0,0}$ [m]	T_p [s]	θ_0 [°]	$\theta_{0,l}$ [°]
ESE	1.51	7.25	67	−22
NNE	2.25	6.75	40	5

Since the NSWE solver needs to be fed with both instantaneous water level and depth-averaged velocity, the spectral data of both baseline tests (Table 1) have been transformed into random free-surface time series using the method described in [53], which consists of the following steps: (i) spectrum discretization into a finite number of frequency intervals; (ii) setting of the wave characteristics of each frequency interval; (iii) summation of all waves (e.g., see [7]). An application of such method is illustrated in Figure 3, where the JONSWAP spectrum (Figure 3a) is that related to the wave characteristics of the NNE forcing (Table 1). The time series of the water-surface level has been generated starting from such spectrum using [53]'s method (Figure 3b).

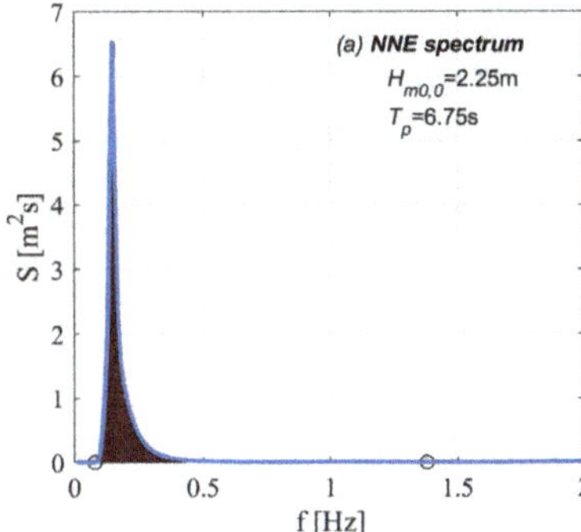

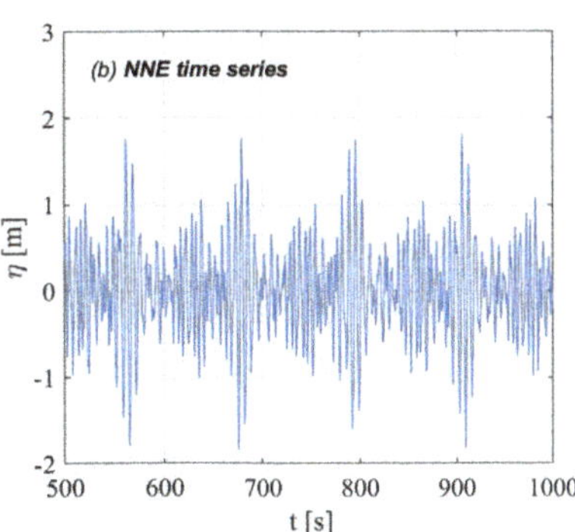

Figure 3. Example of the offshore boundary condition generation: (**a**) spectrum of the NNE wave; (**b**) surface-level time series.

Both numerical simulations have been run using a spatial discretization along the cross-shore and alongshore directions of, respectively, $(\Delta x, \Delta y) = (2, 5)$ m, and a friction coefficient $C_f = 0.005$, consistent with classical coastal processes, e.g., [35].

3. Results

3.1. Step 1: NSP Reconstruction

This section is devoted to present the results obtained from the elaboration of the radar data sequence by means of the NSP strategy described in Section 2. Each considered sequence consists of 256 individual radar images with an interval $\Delta t = 1$ s between two consecutive images and the extent of the pixel in the Cartesian grid is equal to 5 m. An example of both synthetic sea-wave image obtained by means of the NSWE solver (Figure 4a) and amplitude of the corresponding radar image (Figure 4b) is illustrated at the same time instant ($t = 128$ s).

To estimate the local sea-state parameters and bathymetry, the spatial partitioning strategy described in Section 2 is applied to the radar data sequence. In particular, a sub-area size equal to (250×250) m^2 and (300×300) m^2 is adopted for radar sequences belonging to the wave coming from NNE and ESE, respectively. Such difference in the sub-area size is due to the wavelength associated with the directional spectrum evaluated within the whole area investigated by the radar (more details are reported in [31]).

First, the NSP procedure provided the bathymetry reconstructions, which are shown using a pixel spacing of 10 m for both wave types (Figure 5). The reliability of the bathymetric reconstruction using the NSP strategy has been widely demonstrated (e.g., [31]), so the error statistics is reported in Figure 5, for both wave types, only in terms of Root Mean Square Error (RMSE) and correlation coefficient square (R^2), obtained from the comparison with the ground-truth (Figure 2).

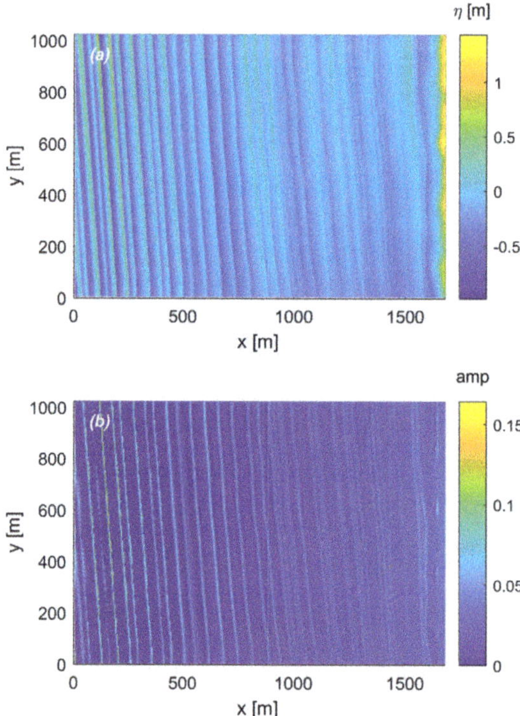

Figure 4. (**a**) Synthetic sea-wave image obtained from the NSWE solver. (**b**) Simulated radar image.

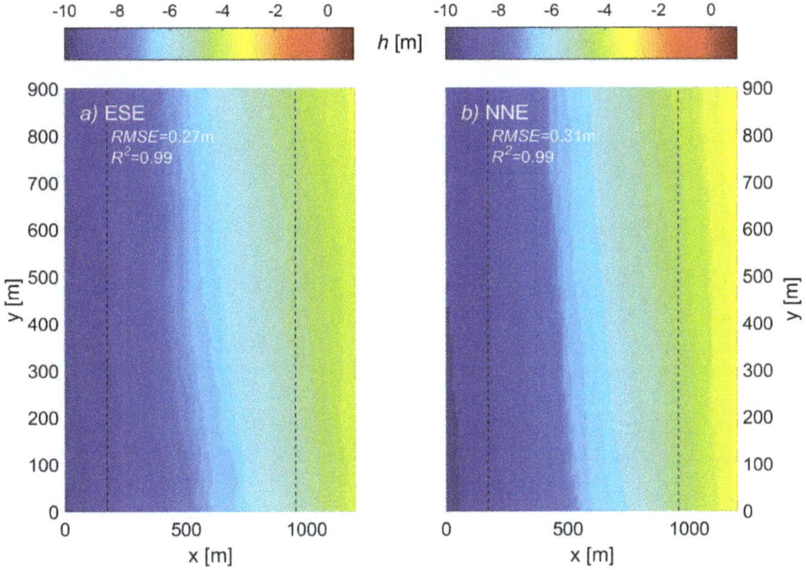

Figure 5. Reconstructed bathymetries: (**a**) ESE and (**b**) NNE simulations. Dashed lines indicate the locations at which the boundary conditions are chosen.

As said, the purpose of this work is to provide suitable beach-inundation predictions starting from the local estimation of the sea-state parameters from the data acquired by means of an X-band wave-radar system. Hence, in the inversion procedure, after application of the BP filter, the following step consists of the application of the MTF to turn from the filtered radar spectrum to the desired sea-wave spectrum $F_w^j(\bar{k}, \omega)$ at each sub-area j. From such spectrum $F_w^j(\bar{k}, \omega)$, it is possible to determine the main sea-state parameters at each sub-area, i.e., the peak period T_p^j, the peak direction θ_p^j, as well as the significant wave height H_{m0}^j. These are used for the generation of the boundary conditions to be applied at specific locations along the cross-shore direction. In particular, these parameters are tied to the spectral moment of zero order m_0 for H_{m0}^j, to the wave number spectrum $F_w^j(\bar{k})$ for θ_p^j and to the one-dimensional frequency spectrum $F_w^j(\omega)$ for T_p^j, by means of the following expressions:

$$m_0^j = \frac{1}{2\pi} \int_k \int_\omega F_w^j(\bar{k}, \omega) d^2\bar{k} d\omega \tag{7}$$

$$F_w^j(\bar{k}) = \int_\omega F_w^j(\bar{k}, \omega) d\omega \tag{8}$$

$$F_w^j(\omega) = \int_{\bar{k}} F_w^j(\bar{k}, \omega) d^2\bar{k} \tag{9}$$

The following figures illustrate the reconstructed maps related to sea-state parameters of both wave types. Specifically, Figure 6 shows the distribution of the significant wave height H_{m0}, which reduces while approaching the shore, this mainly due to the wave breaking effect. On the other hand, the peak wave period slightly changes while moving shoreward (Figure 7). Large and unfeasible changes of T_p occur in shallow waters (i.e., $h < 3$ m), where it is well known that the inversion approach applied to radar sequence data provides unsuitable results.

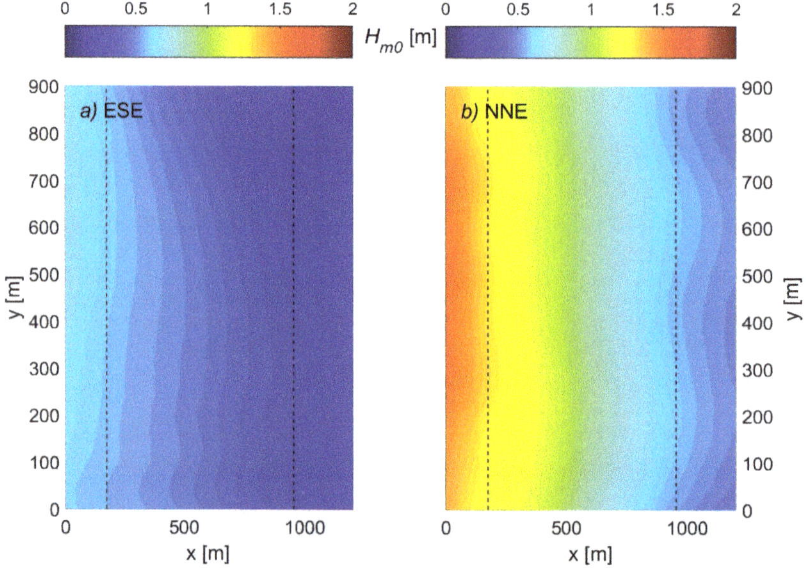

Figure 6. Reconstructed significant wave height: (**a**) ESE and (**b**) NNE simulations. Dashed lines indicate the locations at which the boundary conditions are chosen.

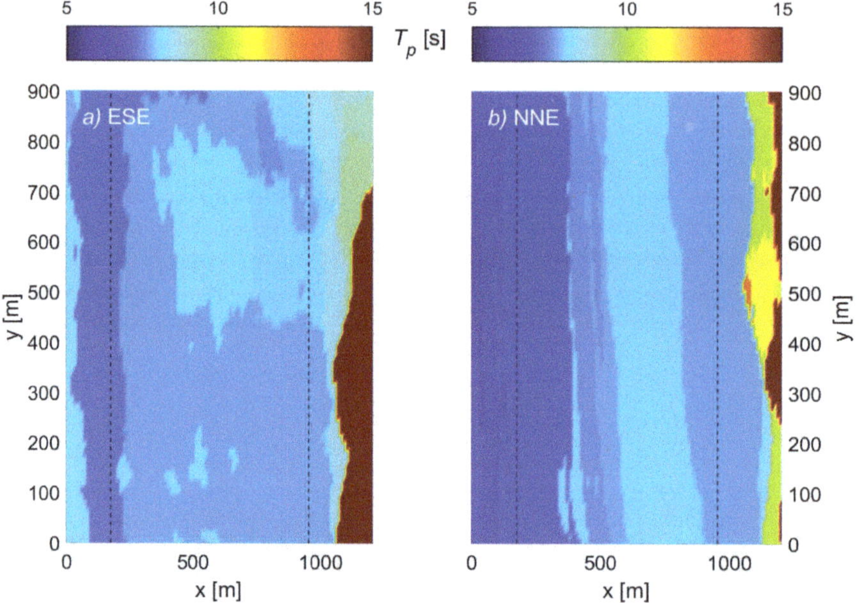

Figure 7. Reconstructed peak periods: (**a**) ESE and (**b**) NNE simulations. Dashed lines indicate the locations at which the boundary conditions are chosen.

Such reconstructions are used to extract, at water depths corresponding to $h_{off} \approx 9$ m and $h_{off} \approx 5$ m (dashed lines in Figures 5–7), the offshore values for the generation of the boundary conditions required by the flood simulations (see Section 3.2).

3.2. Step 2: Boundary and Initial Conditions

The estimate of the wave field is used for the generation of the boundary condition to be applied at specific locations along the cross-shore direction. In particular, simulations have been run starting from water depths of $h_{off} \approx 9$ m, located at a distance $x = 175$ m from the offshore boundary, and of $h_{off} \approx 5$ m, located at $x = 855$ m from the offshore boundary.

A couple of simulations has been run at each chosen depth: one characterized by the waves propagating on a surveyed beach, the other by the waves propagating on an equilibrium-profile bathymetry. While the former type of simulations reproduces the condition for which a beach survey is available, the latter type resembles the case of no availability of bathymetric data. Specifically, it has already been found that both beach inundation and runup on the emerged beach can be well reproduced if a Dean-type equilibrium profile [54] is used instead of the real beach profile. The classical equilibrium profile is expressed by the law $h = A x_{sh}^{2/3}$, where x_{sh} is the distance to shoreline, while A is a site-specific shape parameter. Hence, a suitable equilibrium profile has been built using the following inputs: (i) the bathymetry estimate provided by the NSP method at depths $h > 5$ m, (ii) a constant offshore slope of $1 : 200$, and (iii) a constant value $A = 0.065$ valid throughout the domain (in agreement with previous studies, e.g., [50]).

A total of four flood simulations associated with each baseline test have thus been run (see details in Section 3.3). These are described in Figure 8, where the top panels (a,b,c) refer to an offshore depth $h_{off} \approx 5$ m, the lower panels (d,e,f) to $h_{off} \approx 9$ m. While the left panels (a,d) illustrate the surveyed/ground-truth bathymetries, the middle panels (b,e) show the reconstructed equilibrium-profile bathymetries. The right panels (c,f) show the comparison between ground-truth (solid lines) and equilibrium (dashed lines) profiles.

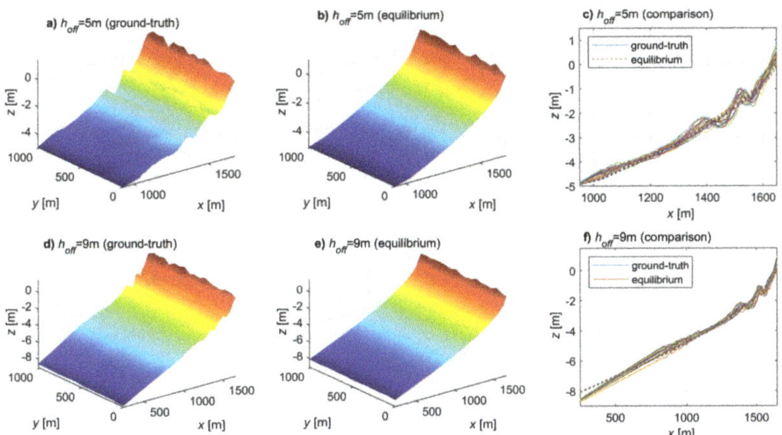

Figure 8. Ground-truth bathymetries (**a**,**d**) and equilibrium-profile-based bathymetries (**b**,**e**) relevant to an offshore depth of 5 m (top panels) and 9 m (bottom panels). Comparisons between ground-truth and equilibrium profiles is also illustrated (**c**,**f**).

The wave characteristics reconstructed using the NSP method (Section 3.1) are all reported in Table 2. These are used for the generation of the boundary conditions required by the flood simulations. It is worth noting that some wave heights reported in Table 2 are smaller than 1 m. Although this value represents a limit for the estimation of sea parameters in the wave-radar frame, the work by [40] showed the robustness of the NSP method for the estimation of the sea-surface current from the radar images acquired in sea conditions of "gentle breeze" with $H_{m0,off} < 1$ m (level 3 on Beaufort scale), a challenging state for the X-band radar. Furthermore, the two proposed scenarios are significantly different each other, this better allowing one to evaluate the potential of the proposed methodology within a wide range of wave characteristics.

Table 2. Wave characteristics used for the flood simulations.

Wave Type	Bathymetry	h_{off} [m]	$H_{m0,off}$ [m]	T_p [s]	$\theta_{0,l}$ [°]
ESE	ground-truth	9.22	0.62	7.35	−17
ESE	equilibrium	9.22	0.62	7.35	−17
ESE	ground-truth	5.34	0.33	7.8	−10
ESE	equilibrium	5.34	0.33	7.8	−10
NNE	ground-truth	9.23	1.32	6.8	0
NNE	equilibrium	9.23	1.32	6.8	0
NNE	ground-truth	5.14	0.64	7.8	0
NNE	equilibrium	5.14	0.64	7.8	0

3.3. Step 3: Flood Simulations

The flood simulations have been run on the bathymetries illustrated in Figure 8 and following the prescriptions of Table 2. The water-surface elevation during the shoreward propagation of the ESE forcing, which is less steep and smaller than the NNE forcing, is illustrated in Figure 9.

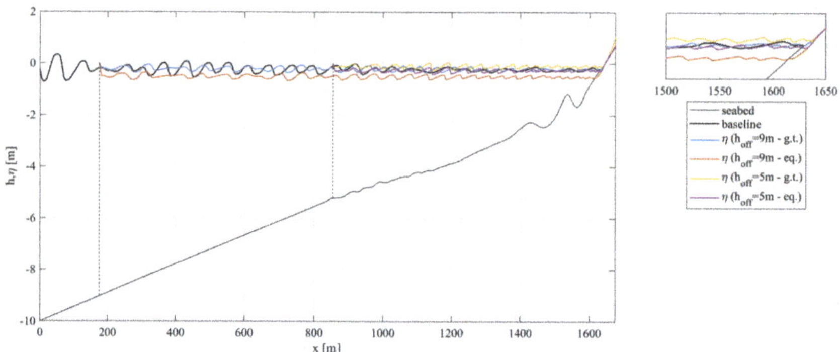

Figure 9. Cross-section of free-surface level at $t = 1000$ s ($\sim$135 waves) for the ESE tests: baseline simulation (black thick line) vs. flood simulations (colored lines). The starting points of the flood simulations, corrsponding to $h \approx 9$ m and $h \approx 5$ m, are also illustrated (vertical dashed lines). The little right panel shows a close-up view in the shallowest region.

During their propagation, the single waves characterizing the spectrum experience several processes, spanning from the shoaling to the breaking, the latter numerically reproduced through a water-surface discontinuity [34]. Here, the comparison between the baseline simulation (black line) and the flood-simulation (colored lines) results at time $t = 1000$ s (corresponding to about 135 waves) is presented. The propagation of the wave trains, though irregular, is consistent and the order of magnitude of the free-surface elevations obtained from the different tests is the same, this being more visible from inspection of the close-up view shown in the right panel. Consistency among the simulations may also be found observing the free-surface evolution throughout the domain (Figure 10). Although some little differences can be noticed between the baseline result (left panels) and the results of the flood tests (middle and right panels), the wave propagation is the same, in terms of: wave direction from the offshore to the inshore, η distribution, wavelength. This also in view of the random nature of the generated series.

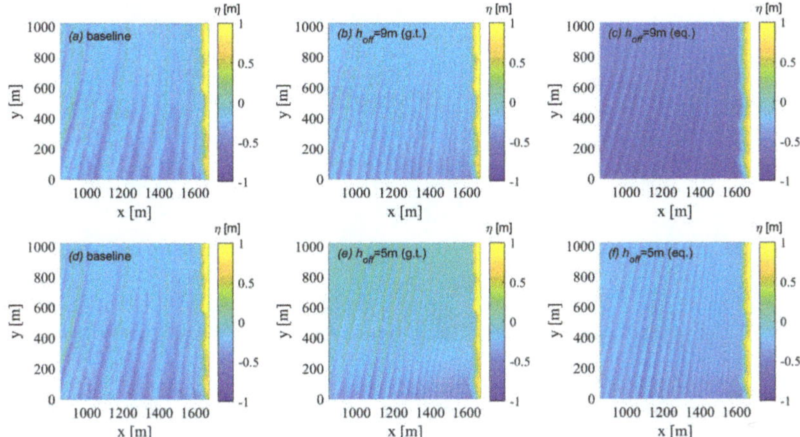

Figure 10. Free-surface level at $t = 1000$ s ($\sim$135 waves) for the ESE tests: baseline simulation (**a,d**), flood simulations with boundary condition at $h \approx 9$ m (**b,c**) and at $h \approx 5$ m (**e,f**). Ground-truth-based (**b,d**) and equilibrium-based (**c,f**) tests.

The NNE wave, steeper and higher than the ESE, also provides good comparisons between baseline and flood simulations. Figures 11 and 12 illustrate the free-surface elevation, which is a bit larger for the baseline condition compared to the flood results, e.g., at the shoreline location (see the right panel of Figure 11). The baseline wave-propagation direction ($\theta_{0,I} = 5°$) is slightly different to that predicted, which is $\theta_{0,I} = 0°$, i.e., waves propagating perpendicularly to the shoreline. The sawtooth shape of the water-surface profile (Figure 11), mainly visible for the baseline simulations (black line), is due both to the wave steepening, induced by the shoaling process, and partially to the offshore displacement of the wave breaking, induced by the absence of dispersion, typical of NSWE models.

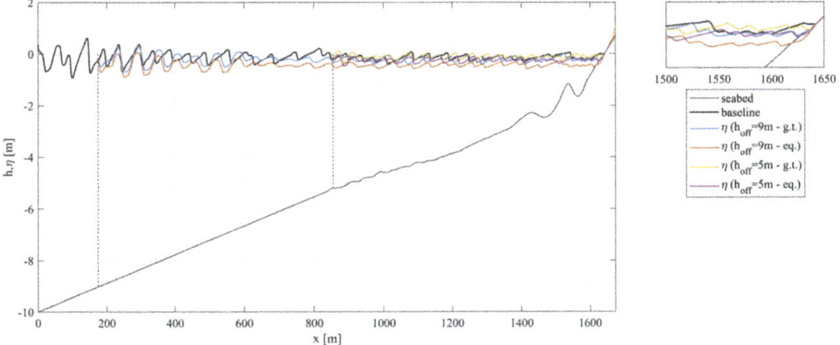

Figure 11. Cross-section of free-surface level at $t = 1000$ s ($\sim$135 waves) for the NNE tests: baseline simulation (black thick line) vs. flood simulations (colored lines). The starting points of the flood simulations, corresponding to $h \approx 9$ m and $h \approx 5$ m, are also illustrated (vertical dashed lines). The little right panel shows a close-up view in the shallowest region.

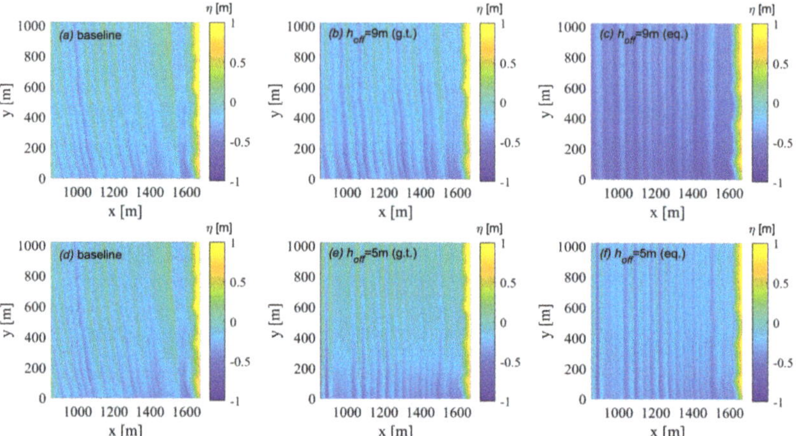

Figure 12. Free-surface level at $t = 1000$ s ($\sim$135 waves) for the NNE tests: baseline simulation (**a,d**), flood simulations with boundary condition at $h \approx 9$ m (**b,c**) and at $h \approx 5$ m (**e,f**). Ground-truth-based (**b,d**) and equilibrium-based (**c,f**) tests.

3.4. Step 4: Beach Inundation

The final step of the proposed methodology consists of the analysis of the beach inundation obtained from the different runs. The maximum shoreline location $x_{s,max}$ observed during the whole runs (test duration: 1000 s) is estimated as the envelope of the shoreline locations obtained at each time step ($\Delta t = 1$ s). For this purpose, every inundation line has been obtained by taking as flooded only the areas characterized by a minimum water depth $d_{flood} = 10$ cm, this being consistent with risk analyses that could be undertaken on the studied zone (e.g., see [7]).

The results of the ESE simulations are illustrated in Figure 13. As can be observed, all colored lines, representing the flood simulations run using either ground-truth or equilibrium-profile beaches, are in very good agreement with the baseline simulation result (black line). These overlaps almost perfectly and the maximum distance is in the order of the grid size ($\Delta x = 2$ m).

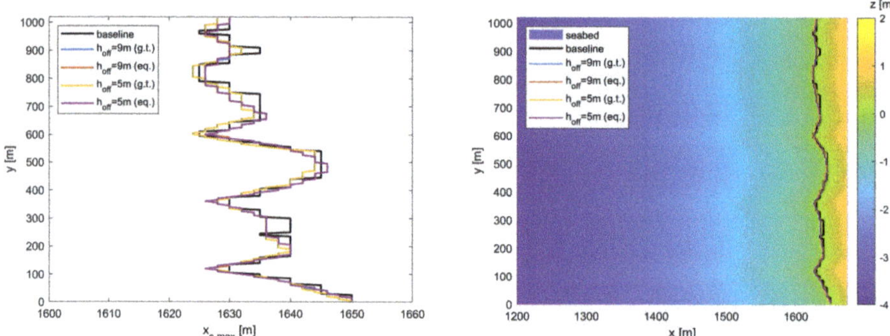

Figure 13. Comparison of max inundation due to the ESE forcing: results of baseline (black line) and flood (colored lines) simulations (**left panel**). The overlap with the original bathymetry is also shown (**right panel**).

The results of the NNE simulations are illustrated in Figure 14, where the beach inundation of the baseline simulation is a bit larger than that obtained from the flood simulations, which tend to slightly underestimate the flooded area. In particular, the region in the range $y = (0–650)$ m provides a good comparison between the two simulation types, with a maximum distance between baseline (black line) and flood-(colored lines) simulation results of 5 m. The upper region, i.e., $y = (650–1000)$ m, shows a weaker, though acceptable, comparison, with a maximum distance of 21 m at $y = 780$ m. This may be attributed to the different propagation direction in the baseline and flood simulations (see Figure 12), due to the slight underestimate of the wave angle, i.e., $\theta_{0,l} = 0°$ (Table 2) versus $\theta_{0,l} = 5°$ (Table 1).

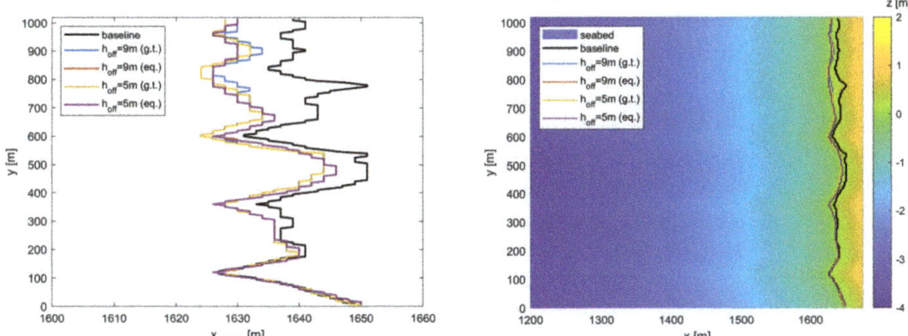

Figure 14. Comparison of max inundation due to the NNE forcing: results of baseline (black line) and flood (colored lines) simulations (**left panel**). The overlap with the original bathymetry is also shown (**right panel**).

In summary, the RMSE values of the flood simulations with respect to the baseline test are reported in Table 3. Errors of the order of the grid size characterize all flood simulations of the ESE case ($RMSE = (2.2–2.3)$ m), while acceptable values are found for the NNE case ($RMSE = (7.6–8.4)$ m). Such results demonstrate the suitability and robustness of the proposed methodology, which combines the use of a remote sensor with the numerical modeling for forecasting and hindcasting purposes related to the beach inundation.

Table 3. RMSE of the $x_{s,max}$ of the flood simulations with respect to the $x_{s,max}$ of the corresponding baseline tests.

Wave Type	Bathymetry	h_{off} [m]	RMSE [m]
ESE	ground-truth	9	2.28
ESE	equilibrium	9	2.34
ESE	ground-truth	5	2.26
ESE	equilibrium	5	2.34
NNE	ground-truth	9	7.67
NNE	equilibrium	9	8.07
NNE	ground-truth	5	8.40
NNE	equilibrium	5	8.05

4. Discussion and Conclusions

The present work aims at improving the skills of typical coastal monitoring systems. Specifically, a methodology is here illustrated based on the combination of a remote sensor (e.g., an X-band marine radar applied to the coastal environment) with a numerical model which solves the nearshore hydrodynamics (e.g., a NSWE solver).

To reach such a purpose and to suitably control the involved variables, the wave field is generated through baseline simulations aimed at reproducing wave conditions representative of the unprotected coastal area of Senigallia (Marche Region, Italy). The tilt modulation is then applied to the wave field generated through each baseline simulation to recreate a real-world radar data sequence, which can be seen as the "initial condition" of the proposed methodology.

The NSP approach is now used to reconstruct both bathymetry and wave field in the region typically covered by an X-band radar, i.e., about (1–6) km from the shoreline. It is demonstrated, here and in previous works (e.g., [31]), that the NSP is able at properly reconstructing directly some of the main variables (seabed depth, wave direction, peak wave period and significant wave height).

Such reconstructed data are then exploited for the generation of both initial and boundary conditions, to be used to feed the NSWE model. The initial condition consists of the reconstructed bathymetry (e.g., referring to seabed depths within 5 m and 9 m) which is extended up to the coast using either an existing survey or an equilibrium-profile-based bathymetry. The reconstructed wave characteristics are used to generate, following [53]'s method, the random time series of free-surface elevation, which characterizes the boundary condition of the flood simulations.

Although the end users of the proposed methodology may choose to run several simulations (e.g., based on the Monte Carlo method, to take into account the randomness of the free-surface boundary condition), the present work aims at illustrating four approaches based on two different boundary depths (around 5 m and 9 m) and two different bathymetric datasets (ground-truth and equilibrium profiles). In each simulation, the waves are propagated from the location where they are reconstructed to the coast. The beach inundation provided by flood simulations of different kinds (i.e., applying the boundary condition at a water depth of either 9 m or 5 m, rather than using either a real-world bathymetry or an equilibrium-profile-derived bathymetry) is compared to that obtained from the baseline simulations, with differences of the order of the used grid size. Better results are those retrieved when a milder forcing, coming from ESE, is used, while steeper and higher waves, coming from NNE, provide a slightly weaker comparison, likely induced by a slight underestimate of the wave direction.

Finally, the present work demonstrates that a suitable coupling between NSP approach and numerical simulations leads to good results in terms of coastal flooding in real-world environments. This can provide useful information for the coastal environment and the related stakeholders (coast authorities, municipalities, beach resort owners, etc.), mainly dealing with: (i) real-time predictions of coastal flooding and warning systems; (ii) hazard mapping and coastal risk analysis; (iii) integrated coastal zone management.

A real-world application of the proposed methodology is expected in the near future, based on the NSP application to a radar data sequence, combined to depth-averaged/shallow-water simulations.

Author Contributions: Conceptualization, M.P. and G.L.; methodology, M.P.; validation, M.P. and G.L.; investigation, M.P. and G.L.; resources, M.P. and G.L.; writing—original draft preparation, M.P.; writing—review and editing, M.P. and G.L.

Funding: This research was partially funded by the Office of Naval Research Global (UK), through the MORSE Project (Research Grant N62909-17-1-2148) .

Acknowledgments: The authors would like to thank the municipality of Senigallia for sharing the bathymetric survey.

References

1. CCCuk. UK Climate Change Risk Assessment 2017. Synthesis report: Priorities for the next five years. In *UK Climate Change Risk Assessment 2017 Synthesis Report of Committee on Climate Change*; Technical Report; CCCuk: London, UK, 2016.
2. Barcikowska, M.J.; Weaver, S.J.; Feser, F.; Russo, S.; Schenk, F.; Stone, D.A.; Wehner, M.F.; Zahn, M. Euro-Atlantic winter storminess and precipitation extremes under 1.5 °C vs. 2 °C warming scenarios. *Earth Syst. Dyn.* **2018**, *9*, 679–699. [CrossRef]
3. Lin, N.; Emanuel, K.A.; Smith, J.A.; Vanmarcke, E. Risk assessment of hurricane storm surge for New York City. *J. Geophys. Res. Atmos.* **2010**, *115*. [CrossRef]

4.   Bosom García, E.; Jiménez Quintana, J.A. Probabilistic coastal vulnerability assessment to storms at regional scale: application to Catalan beaches (NW Mediterranean). *Nat. Hazards Earth Syst. Sci.* **2011**, *11*, 475–484. [CrossRef]

5.   Perini, L.; Calabrese, L.; Salerno, G.; Ciavola, P.; Armaroli, C. Evaluation of coastal vulnerability to flooding: comparison of two different methodologies adopted by the Emilia-Romagna region (Italy). *Nat. Hazards Earth Syst. Sci.* **2016**, *16*, 181–194. [CrossRef]

6.   Leo, F.D.; Besio, G.; Zolezzi, G.; Bezzi, M. Coastal vulnerability assessment: through regional to local downscaling of wave characteristics along the Bay of Lalzit (Albania). *Nat. Hazards Earth Syst. Sci.* **2019**, *19*, 287–298. [CrossRef]

7.   Postacchini, M.; Lalli, F.; Memmola, F.; Bruschi, A.; Bellafiore, D.; Lisi, I.; Zitti, G.; Brocchini, M. A model chain approach for coastal inundation: Application to the bay of Alghero. *Estuar. Coast. Shelf Sci.* **2019**, *219*, 56–70. [CrossRef]

8.   Neumann, B.; Vafeidis, A.T.; Zimmermann, J.; Nicholls, R.J. Future coastal population growth and exposure to sea-level rise and coastal flooding—A global assessment. *PLoS ONE* **2015**, *10*, e0118571. [CrossRef] [PubMed]

9.   UNEP/MAP. *Protocol on Integrated Coastal Zone Management in the Mediterranean;* Technical Report; UNEP/MAP: Nairobi, Kenya, 2008.

10.   Rochette, J.; Bille, R. ICZM protocols to regional seas conventions: What? why? how? *Mar. Policy* **2012**, *36*, 977–984. [CrossRef]

11.   Ernoul, L.; Wardell-Johnson, A. Environmental discourses: Understanding the implications on ICZM protocol implementation in two Mediterranean deltas. *Ocean Coast. Manag.* **2015**, *103*, 97–108. [CrossRef]

12.   Herbers, T.; Jessen, P.; Janssen, T.; Colbert, D.; MacMahan, J. Observing ocean surface waves with GPS-tracked buoys. *J. Atmos. Ocean. Technol.* **2012**, *29*, 944–959. [CrossRef]

13.   Ohlmann, J.C.; Fewings, M.R.; Melton, C. Lagrangian observations of inner-shelf motions in Southern California: Can surface waves decelerate shoreward-moving drifters just outside the surf zone? *J. Phys. Oceanogr.* **2012**, *42*, 1313–1326. [CrossRef]

14.   Rogers, W.E.; Holland, K.T. A study of dissipation of wind-waves by mud at Cassino Beach, Brazil: Prediction and inversion. *Cont. Shelf Res.* **2009**, *29*, 676–690. [CrossRef]

15.   Brocchini, M.; Calantoni, J.; Postacchini, M.; Sheremet, A.; Staples, T.; Smith, J.; Reed, A.H.; Braithwaite, E.F., III; Lorenzoni, C.; Russo, A.; et al. Comparison between the wintertime and summertime dynamics of the Misa River estuary. *Mar. Geol.* **2017**, *385*, 27–40. [CrossRef]

16.   Melville, W.; Loewen, M.R.; Felizardo, F.C.; Jessup, A.T.; Buckingham, M. Acoustic and microwave signatures of breaking waves. *Nature* **1988**, *336*, 54. [CrossRef]

17.   Nieto Borge, J.; Rodríguez, G.R.; Hessner, K.; González, P.I. Inversion of marine radar images for surface wave analysis. *J. Atmos. Ocean. Technol.* **2004**, *21*, 1291–1300. [CrossRef]

18.   Archetti, R.; Paci, A.; Carniel, S.; Bonaldo, D. Optimal index related to the shoreline dynamics during a storm: the case of Jesolo beach. *Nat. Hazards Earth Syst. Sci.* **2016**, *16*, 1107–1122. [CrossRef]

19.   Benetazzo, A.; Serafino, F.; Bergamasco, F.; Ludeno, G.; Ardhuin, F.; Sutherland, P.; Sclavo, M.; Barbariol, F. Stereo imaging and X-band radar wave data fusion: An assessment. *Ocean Eng.* **2018**, *152*, 346–352. [CrossRef]

20.   Díaz, H.; Catalán, P.; Wilson, G. Quantification of Two-Dimensional Wave Breaking Dissipation in the Surf Zone from Remote Sensing Data. *Remote Sens.* **2018**, *10*, 38.

21.   Bell, P.S.; Osler, J.C. Mapping bathymetry using X-band marine radar data recorded from a moving vessel. *Ocean. Dyn.* **2011**, *61*, 2141–2156. [CrossRef]

22.   Ludeno, G.; Orlandi, A.; Lugni, C.; Brandini, C.; Soldovieri, F.; Serafino, F. X-band marine radar system for high-speed navigation purposes: A test case on a cruise ship. *IEEE Geosci. Remote Sens. Lett.* **2014**, *11*, 244–248. [CrossRef]

23.   Lund, B.; Graber, H.C.; Hessner, K.; Williams, N.J. On shipboard marine X-band radar near-surface current "calibration". *J. Atmos. Ocean. Technol.* **2015**, *32*, 1928–1944. [CrossRef]

24.   Ludeno, G.; Brandini, C.; Lugni, C.; Arturi, D.; Natale, A.; Soldovieri, F.; Gozzini, B.; Serafino, F. Remocean system for the detection of the reflected waves from the costa concordia ship wreck. *IEEE J. Sel. Top. Appl. Earth Obs. Remote Sens.* **2014**, *7*, 3011–3018. [CrossRef]

25. Ludeno, G.; Reale, F.; Dentale, F.; Carratelli, E.; Natale, A.; Soldovieri, F.; Serafino, F. An X-band radar system for bathymetry and wave field analysis in a harbour area. *Sensors* **2015**, *15*, 1691–1707. [CrossRef] [PubMed]
26. Gangeskar, R. Ocean current estimated from X-band radar sea surface, images. *IEEE Trans. Geosci. Remote. Sens.* **2002**, *40*, 783–792. [CrossRef]
27. Senet, C.M.; Seemann, J.; Flampouris, S.; Ziemer, F. Determination of bathymetric and current maps by the method DiSC based on the analysis of nautical X-band radar image sequences of the sea surface (November 2007). *IEEE Trans. Geosci. Remote Sens.* **2008**, *46*, 2267–2279. [CrossRef]
28. Young, I.R.; Rosenthal, W.; Ziemer, F. A three-dimensional analysis of marine radar images for the determination of ocean wave directionality and surface currents. *J. Geophys. Res. Oceans* **1985**, *90*, 1049–1059. [CrossRef]
29. Senet, C.M.; Seemann, J.; Ziemer, F. The near-surface current velocity determined from image sequences of the sea surface. *IEEE Trans. Geosci. Remote Sens.* **2001**, *39*, 492–505. [CrossRef]
30. Serafino, F.; Lugni, C.; Soldovieri, F. A novel strategy for the surface current determination from marine X-band radar data. *IEEE Geosci. Remote Sens. Lett.* **2010**, *7*, 231–235. [CrossRef]
31. Ludeno, G.; Postacchini, M.; Natale, A.; Brocchini, M.; Lugni, C.; Soldovieri, F.; Serafino, F. Normalized Scalar Product Approach for Nearshore Bathymetric Estimation From X-Band Radar Images: An Assessment Based on Simulated and Measured Data. *IEEE J. Ocean. Eng.* **2018**, *43*, 221–237. [CrossRef]
32. Bellafiore, D.; Zaggia, L.; Broglia, R.; Ferrarin, C.; Barbariol, F.; Zaghi, S.; Lorenzetti, G.; Manfè, G.; De Pascalis, F.; Benetazzo, A. Modeling ship-induced waves in shallow water systems: The Venice experiment. *Ocean Eng.* **2018**, *155*, 227–239. [CrossRef]
33. Gaeta, M.; Bonaldo, D.; Samaras, A.; Carniel, S.; Archetti, R. Coupled Wave-2D Hydrodynamics Modeling at the Reno River Mouth (Italy) under Climate Change Scenarios. *Water* **2018**, *10*, 1380. [CrossRef]
34. Brocchini, M.; Bernetti, R.; Mancinelli, A.; Albertini, G. An efficient solver for nearshore flows based on the WAF method. *Coast. Eng.* **2001**, *43*, 105–129. [CrossRef]
35. Briganti, R.; Torres-Freyermuth, A.; Baldock, T.E.; Brocchini, M.; Dodd, N.; Hsu, T.J.; Jiang, Z.; Kim, Y.; Pintado-Patiño, J.C.; Postacchini, M. Advances in numerical modelling of swash zone dynamics. *Coast. Eng.* **2016**, *115*, 26–41. [CrossRef]
36. Kennedy, A.B.; Chen, Q.; Kirby, J.T.; Dalrymple, R.A. Boussinesq modeling of wave transformation, breaking, and runup. I: 1D. *J. Waterw. Port Coast. Ocean Eng.* **2000**, *126*, 39–47. [CrossRef]
37. Antuono, M.; Colicchio, G.; Lugni, C.; Greco, M.; Brocchini, M. A depth semi-averaged model for coastal dynamics. *Phys. Fluids* **2017**, *29*, 056603. [CrossRef]
38. Tonelli, M.; Petti, M. Hybrid finite volume—Finite difference scheme for 2DH improved Boussinesq equations. *Coast. Eng.* **2009**, *56*, 609–620. [CrossRef]
39. Shi, F.; Kirby, J.T.; Harris, J.C.; Geiman, J.D.; Grilli, S.T. A high-order adaptive time-stepping TVD solver for Boussinesq modeling of breaking waves and coastal inundation. *Ocean Model.* **2012**, *43*, 36–51. [CrossRef]
40. Ludeno, G.; Nasello, C.; Raffa, F.; Ciraolo, G.; Soldovieri, F.; Serafino, F. A comparison between drifter and X-band wave radar for sea surface current estimation. *Remote Sens.* **2016**, *8*, 695. [CrossRef]
41. Plant, W. Studies of backscattered sea return with a CW, dual-frequency, X-band radar. *IEEE J. Ocean. Eng.* **1977**, *2*, 28–36. [CrossRef]
42. Lee, P.; Barter, J.; Beach, K.; Hindman, C.; Lake, B.; Rungaldier, H.; Shelton, J.; Williams, A.; Yee, R.; Yuen, H. X band microwave backscattering from ocean waves. *J. Geophys. Res. Oceans* **1995**, *100*, 2591–2611. [CrossRef]
43. Wetzel, L.B. Electromagnetic scattering from the sea at low grazing angles. In *Surface Waves and Fluxes*; Springer: Dordrecht, The Netherlands, 1990; pp. 109–171.
44. Hessner, K.; Reichert, K.; Borge, J.C.N.; Stevens, C.L.; Smith, M.J. High-resolution X-band radar measurements of currents, bathymetry and sea state in highly inhomogeneous coastal areas. *Ocean Dyn.* **2014**, *64*, 989–998. [CrossRef]
45. Ludeno, G.; Flampouris, S.; Lugni, C.; Soldovieri, F.; Serafino, F. A novel approach based on marine radar data analysis for high-resolution bathymetry map generation. *IEEE Geosci. Remote Sens. Lett.* **2014**, *11*, 234–238. [CrossRef]
46. Raffa, F.; Ludeno, G.; Patti, B.; Soldovieri, F.; Mazzola, S.; Serafino, F. X-band wave radar for coastal upwelling detection off the southern coast of Sicily. *J. Atmos. Ocean. Technol.* **2017**, *34*, 21–31. [CrossRef]
47. Lund, B.; Collins, C.O.; Graber, H.C.; Terrill, E.; Herbers, T.H. Marine radar ocean wave retrieval's dependency on range and azimuth. *Ocean Dyn.* **2014**, *64*, 999–1018. [CrossRef]

48. Kriebel, D.L.; Dean, R.G. Convolution method for time-dependent beach-profile response. *J. Waterw. Port Coast. Ocean Eng.* **1993**, *119*, 204–226. [CrossRef]
49. Soldini, L.; Antuono, M.; Brocchini, M. Numerical modeling of the influence of the beach profile on wave run-up. *J. Waterw. Port Coast. Ocean Eng.* **2013**, *139*, 61–71. [CrossRef]
50. Postacchini, M.; Soldini, L.; Lorenzoni, C.; Mancinelli, A. Medium-term dynamics of a middle Adriatic barred beach. *Ocean Sci.* **2017**, *13*, 719. [CrossRef]
51. Parlagreco, L.; Melito, L.; Devoti, S.; Perugini, E.; Soldini, L.; Zitti, G.; Brocchini, M. Monitoring for Coastal Resilience: Preliminary Data from Five Italian Sandy Beaches. *Sensors* **2019**, *19*, 1854. [CrossRef]
52. Melito, L.; Postacchini, M.; Sheremet, A.; Calantoni, J.; Zitti, G.; Darvini, G.; Brocchini, M. Hydrodynamics at a Microtidal Inlet: Analysis of Propagation of the Main Wave Components. *Estuar. Coast. Shelf Sci.* **2019**, under review.
53. Liu, Z.; Frigaard, P. *Generation and Analysis of Random Waves*; Technical Report; Aalborg Universitet: Aalborg, Denmark, 1990.
54. Dean, R.G. Equilibrium beach profiles: Characteristics and applications. *J. Coast. Res.* **1991**, *7*, 53–84.

Journal of
*Marine Science
and Engineering*

MDPI

Article

The Influence of Free Long Wave Generation on the Shoaling of Forced Infragravity Waves

Theo Moura [1,2] and Tom E. Baldock [3,*]

[1]   Hidromod ltd., Porto Salvo, Lisbon, 2740-278 Lisbon, Portugal
[2]   Centre for Marine and Environmental Research, University of Algarve, 8005-139 Faro, Portugal
[3]   School of Civil Engineering, University of Queensland, St Lucia QLD 4072, Australia
*   Correspondence: t.baldock@uq.edu.au

Received: 9 August 2019; Accepted: 30 August 2019; Published: 4 September 2019

Abstract: Different conceptual models for forced infragravity (long) waves exist in the literature, which suggest different models for the behavior of shoaling forced waves and the possible radiation of free long waves in that process. These are discussed in terms of existing literature. A simple numerical model is built to evaluate the wave shape in space and time during shoaling of forced waves with concurrent radiation of free long waves to ensure mass continuity. The same qualitative results were found when performing simulations with the COULWAVE model using the radiation stress term in the momentum equation to force the generation and propagation of bound waves. Both model results indicate a strong frequency dependence in the shoaling rate and on the lag of the total long wave with respect to the forcing, consistent with observations in the literature and more complex evolution models. In this approach, a lag of the long wave is only observed in the time domain, not in the space domain. In addition the COULWAVE is used to investigate dissipation rates of incident free and forced long waves inside the surf zone. The results also show a strong frequency dependence, as previously suggested in the literature, which can contribute to the total rate of decay of the incident forced wave after short wave breaking.

Keywords: Long (infragravity) waves; forced waves; bound waves; shoaling; dissipation

---

1. Introduction

The importance of forced long waves, or bound waves, to the oceans dynamics is well established in the literature, especially in shallow waters. However, in nature, some processes related to their generation, propagation, shoaling and dissipation are yet not fully understood, with eventual conflicting concepts to explain the same processes. A plausible partial explanation is that those concepts emerge from different interpretations of the bound wave theory first introduced by Longuet-Higgins and Stewart [1]. For steady groups, two solutions have been proposed for forced waves by wave groups, which differ in the definition of the water level about which the forced waves oscillate. These indicate a forced wave oscillating either around or solely below the mean water level for the short waves, respectively (Figure 1). The difference between the two interpretations is the constant of integration, which can be taken as the still water level or mean water level under the group, which for a steady condition may be arbitrarily chosen giving the long wave response as [1]

$$\overline{\eta} = -\frac{S_{xx}}{\rho\left(gh - c_g^2\right)} + constant \tag{1}$$

and which has been widely adopted to estimate the forced wave amplitude from the magnitude of the radiation stress (S_{xx}) derived from the short wave amplitude or wave envelope. Here, ρ is the constant water density, c_g is the group velocity, g is the acceleration due to gravity and h is the local water depth.

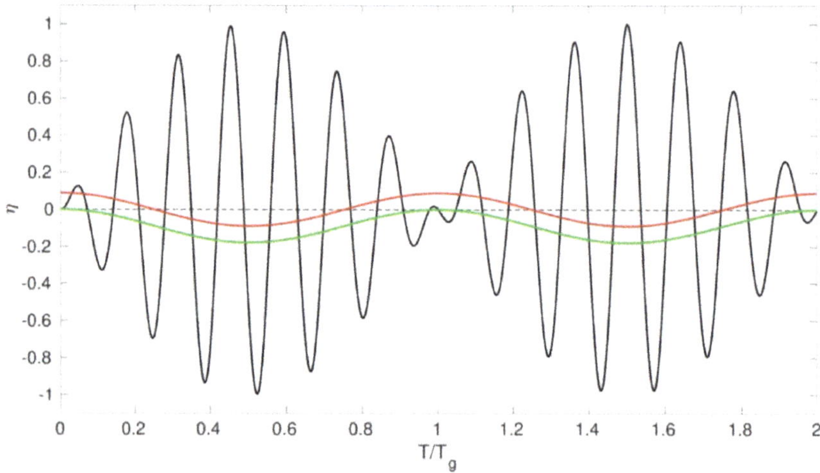

Figure 1. Two different graphical interpretations of the bound wave forced by a short wave groups (black line) by Longuet-Higgins and Stewart [1] with positive and negative elevations (red line), and Mei [2] with purely negative elevation (green line).

This solution emerges from the mass and momentum balance equations including the radiation stress forcing function as

$$\frac{\partial \bar{\eta}}{\partial t} + \frac{\partial (h\bar{u})}{\partial x} = 0 \tag{2}$$

$$\frac{\partial \bar{u}}{\partial t} + g\frac{\partial \bar{\eta}}{\partial x} = -\frac{1}{\rho h}\frac{\partial S_{xx}}{\partial x}. \tag{3}$$

where $\bar{\eta}$ and $\bar{u}$ are the long wave surface displacement and horizontal velocity, respectively. The solution domain is from the bed to the free surface, and initial conditions are discussed further below.

Equation (1) is the forced part of solution to the non-homogenous differential equations, which allows free waves (the homogenous solution) to be added to the forced wave term [3]. The additional free waves may become important in a transient solution, but these have most often been ignored in analysis and modelling of bound wave propagation by treating this as a quasi-steady process. Transient solutions to similar problems dates back to Proudman [4], who proposed dynamical theories of water motions induced by a moving pressure field. Whitham [5] first addressed this issue with respect to the bound wave problem. Later, Molin [6] and Mei and Benmoussa [7] extended the investigation to sloping bottoms. The difference in interpretation of the forced wave as illustrated in Figure 1 is important during shoaling. With a bound wave with both positive and negative components then shoaling (an increase in amplitude in shallower water) does not necessarily require radiation of free waves to ensure continuity of volume, but with a bound wave of pure depression, then an increase in amplitude of that depression requires radiation of free waves to ensure continuity. Radiation of those free waves then changes the shape of the total infragravity wave during shoaling, and may also influence the commonly observed lag between incident bound waves and the incoming short wave envelope.

Nielsen and Baldock [8] presented simple analytical transient solutions to the response of the surface elevation to both non-resonant (finite water depth conditions) and resonant (shallow water conditions) radiation stress forcing. This solution has the same initial conditions as those proposed for startup of a submarine landslide that may trigger tsunami [9,10], i.e., a flat water surface and no flow velocity. In summary, for a 1-D scenario and assuming an abrupt onset of a non-resonant positive forcing with a Gaussian shape, three waves are generated: a purely negative forced wave, propagating phase locked with the forcing, and two free waves with positive amplitude and opposite direction of propagation. While the shape of all three waves are the same as the forcing, the free wave amplitudes depend on the ratio between the wave group velocity and the shallow water wave celerity as

$$\left(A_{free}^{+}, A_{free}^{-}\right) = \left(-\frac{A_{forced}}{2}\left[1 + \frac{c_g}{\sqrt{gh}}\right], -\frac{A_{forced}}{2}\left[1 - \frac{c_g}{\sqrt{gh}}\right]\right) \tag{4}$$

where A_{forced}, A_{free}^{+} and A_{free}^{-} are the amplitudes of the forced and the two free waves respectively.

At the resonant condition (short waves in shallow water), the free long waves arising from the initial condition cannot radiate away, forming a N-shaped wave, with a positive hump leading the forced wave. The total long wave surface grows linearly in time, but with the shape of the derivative of the forcing, rather than the inverse shape as in the steady solution. Changes in the shape of the bound wave from a symmetrical depression to an asymmetrical depression plus surge leading wave has been reported to occur during the shoaling processes, where again purely forced [11] and a mix of forced and free waves [8] interpretations have been provided.

During short wave breaking, the assumption that bound waves are released as free waves is commonly adopted [12–14]. However, strong infragravity wave dissipation has also been observed, suggesting that bound waves are not released during the breaking process, but remain phase locked (forced) in the surf zone decaying with the short wave forcing [15–17]. Breaking of shorter infragravity waves close to the shoreline [18], friction [19] and energy transfer [20] and dissipation due to the existence of turbulence in the surf zone [13] have been proposed or suggested as additional or alternate reasons for dissipation of forced wave energy.

Even though the generation of free waves due to the unsteady nature of forced bound waves have been discussed in different studies [7,8,21], it is not clearly addressed in others, where the interpretation of the results relies on the quasi-steady solution [11–13,22]. To date, generation and radiation of free waves during bound wave shoaling has not been documented in observations, but the shape of the long wave and its phase (lag) relative to the forcing is observed to change. In part this may be due to errors in wave generation, the predominance of shallow water conditions, and the likely small magnitude of the free waves. Nevertheless, it is important to comprehend the possible effects of such free wave generation in the interpretation of propagating and shoaling infragravity waves. The analysis is based on numerical results and previous published laboratory data.

The paper is organised as follows: In Section 2, aspects related to the shape change of the bound wave during shoaling, the possible consequences of the generation of free waves on the observed shoaling rate and lag of the long wave are discussed. A simple semi-analytical model is constructed to evaluate the model concepts of Nagase and Mizuguchi [21] and Nielsen and Baldock [8] and to quantify how the group frequency influences the apparent shoaling rate and lag of the long waves. In Section 3, the possibility of dissipation of long waves by surf zone turbulence is investigated using the COULWAVE Boussinesq wave model, and particularly if the frequency dependence suggested by Battjes [13] occurs in the model results. Final remarks are presented in Section 4.

2. Bound Wave Shoaling

During bound wave shoaling two processes have often been observed: changes in shape, with the development of a leading positive surge and a lag of the long wave in time relative to the short wave envelope; and a shoaling rate dependence on the slope and frequency of the bound wave [12,13,18,23,24]. From the static quasi equilibrium approach, the positive surge ahead of the short wave group can be interpreted as the stronger response of surface elevation to the radiation stress gradients in shallow water and an asymmetry in the short wave envelope as the wave group shoals. The main argument presented by Baldock [11] when analysing single transient wave groups was the very close match of the spatial gradients of the wave envelope (radiation stress) with the spatial gradients of the long, as in Longuet-Higgins and Stewart [1]. A similar observation from measured data is presented in Figure 2, but for an individual wave group from a random wave case from Baldock and Huntley [25]. The lag of the long wave relative to the short wave envelope is also very small in this case, but appears larger in the time domain than in the space domain. It is important to note that these data were obtained with 2nd order generation, which minimises the generation of free (error) waves at the wavemaker (see Orzaghova et al., [26] for other examples).

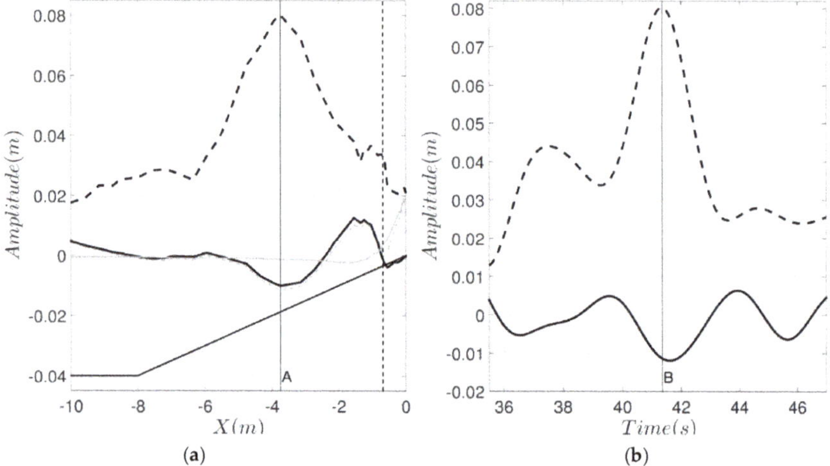

Figure 2. (**a**) Spatial visualization of short wave envelope (dashed black line), total long wave surface elevation (grey dashed), long wave with setup removed (full black line) and mean setup (full grey line), over bottom topography shown (solid black line). (**b**) Time evolution of short wave envelope (dashed line) and total long wave (full line) at x = −3.75 m (location A in left panel). Random wave case J6033A from Baldock and Huntley [25].

The relationship between the wave envelope and the bound wave is commonly analysed based on time series from a point measurement. This is despite the fact that because the propagating wave group is not steady and not symmetric during shoaling the wave shape measured in the time domain is not the same as the instantaneous shape observed in the space domain, see direct comparisons by Baldock [11] and Figure 2 below. In the left panel in Figure 2 the long wave is in antiphase with the forcing and both are asymmetric. In the right panel, no larger leading wave is observed, even though this is present in the left panel, which means that wave only is generated or amplified very close to the shore, in shallow water. With measurements in the time domain, lags between the forcing and bound wave are often reported [22,27,28]. That occurs even when no lag is observed in the spatial domain (Figure 2).

In a time series analysis, the bound wave lag is often associated with the intensity of energy transfer from the forcing to the long wave, where a greater lag means more energy transfer. For the same beach slope, higher frequency bound waves appear to lag more than lower frequency bound waves, resulting in higher shoaling rates at higher frequencies. This difference in shoaling rates is clearly evident in data from Baldock et al. [24] and Janssen et al. [12]. Battjes et al. [13] showed that this shoaling rate is determined by the wave group length relative to the bottom slope, a normalised beach slope, varying from approximately $\sim h^{-\frac{1}{4}}$ (free wave shoaling rate) for relatively lower frequencies to $\sim h^{-\frac{5}{2}}$ (steady solution shoaling rate) for relatively higher frequencies. It is noted that the shoaling at higher frequencies corresponds to the amplitude growth proposed for the quasi-steady solution of Longuet-Higgins and Stewart [1,29], despite no energy transfer or lag being proposed at that time.

In the literature, two terms in the energy balance equation are relevant for modelling the energy exchanges between short and long waves. To the authors' understanding there are still some difficulties in the interpretation of their role in the energy transfer from short to bound waves. The problem has been introduced by Longuet-Higgins and Stewart [1,29], followed by Whitham [5], who attempted to clarify the issue addressed by the previous authors. In the full energy balance introduced by Philips [30] and Schaffer [3], and using their notation for clarity, the conservation of total energy is given by

$$\frac{\partial E}{\partial t} + \frac{\partial W}{\partial x} = 0 \tag{5a}$$

where E is the total energy density and W is the total energy flux. The short wave energy equation is

$$\frac{\partial}{\partial t}\left\{E_s - \frac{1}{2}\rho(h+\eta)U_s^2\right\} + \frac{\partial}{\partial}\left\{U_c E_s + W_s - \frac{1}{2}\rho(h+\eta)UU_s^2\right\} + S_{xx}\frac{\partial U_c}{\partial x} - U_s\frac{\partial S_{xx}}{\partial x} = 0 \tag{5b}$$

where subscript s and c denote the short and long waves (current) respectively. The equation for the free slow varying current or long wave is

$$\frac{\partial E_c}{\partial t} + \frac{\partial W_c}{\partial x} + U\frac{\partial S_{xx}}{\partial x} = 0 \tag{5c}$$

The two terms related to energy transfer between short wave (via S_{xx}) and long waves, with U as the bound long wave velocity (equivalent to $\overline{u}$ in (2) and (3) are

$$\frac{\partial S_{xx}}{\partial x} \, U\frac{\partial S_{xx}}{\partial x}, \tag{6}$$

$$S_{xx}\frac{\partial U}{\partial x} \tag{7}$$

While Battjes et al. [13] have considered (6) as the main driver of energy transfer, Henderson et al. [31], Guedes et al. [32] and others considered (7) as the principal mechanism. Schaffer [3] (see page 564), taking in consideration free and forced long waves, suggested (6) as being the interaction between S_{xx} and the forced wave current, and (7) as the interaction between S_{xx} and the free current (free long waves). When considering the short wave and free current (tide) interaction (7) was the only term considered by Longuet-Higgins and Stewart [29], with the forced bound wave being a product of that interaction.

Janssen et al. [12], following the work of Bowers [33] and Van Leeuwen [34], presented analytical and numerical solutions for the bound wave amplitude and phase shift induced by changes in depth. Battjes et al. [13], assuming quasi-steady waves, derived a phase-average rate of energy transfer based on the phase lag between the wave envelope and the forced wave ($\Delta\phi$ - the deviation from the equilibrium solution, π), the amplitudes of radiation stress ($\hat{S}$) and the forced long wave velocity ($\hat{U}$)

$$R \approx U\frac{\partial S_{xx}}{\partial x} \cong \frac{1}{2}\kappa\hat{U}(f)\hat{S}(f)\sin\Delta\phi \tag{8}$$

where $\kappa = 2\pi f/c_g$ is the forced wave number, at individual frequencies, f, and (ˆ) denotes real amplitudes. Good agreement between this model and laboratory data was found.

Generally, the discussion regarding bound wave shoaling requires the assumption of steady wave conditions, even though the possibility of free wave generation during the shoaling process is recognized in some studies. With a dynamic process, Nagase and Mizuguchi [21] suggested that the observed smaller growth rate compared to the steady solution of Longuet-Higgins and Stewart [1] of the bound wave is a consequence of the superposition of forced and free waves generated due to the transient behaviour of the bound (forced) wave on the slope, which is similar to the suggestion of Nielsen and Baldock [8]. Based on this assumption, even if the shoaling of the forced wave part is independent of the wave group frequency and follows the equilibrium solution, due to the superposition of forced and free waves, an apparent distinct rate of energy transfer (as defined by the rate of change of the total wave amplitude) would exist for wave groups with different length and the same group velocity, c_g. The apparent lag of the negative pulse would also depend on the group frequency.

Using the conceptual models of Nagase and Mizuguchi and Nielsen and Baldock, it is straightforward to calculate the evolution of the total long wave expected from superposition of propagating forced wave and free long waves, as illustrated in Figure 3. Consider a Gaussian-shaped forced wave with amplitude A and length (L_{scale}) with its center location at x_i

$$\eta = -Ae^{\left(\frac{x_i-x}{L_{scale}}\right)},\tag{9}$$

propagating with constant form over a slope, and with an increasing amplitude proportional to $x \sim h^{-5/2}$ (a shoaling rate proportional to the equilibrium solution), where x is the horizontal distance. At each new time step, the gain in amplitude by the bound wave (an increasing negative depression) is added to the forward free wave as a positive elevation with same shape as (9), which is shoaling also as a free wave ($x \sim h^{-1/4}$), to ensure continuity (1). The Gaussian depression is traveling at slower velocity, at the wave group velocity c_g, which is defined in terms of the short wave period T and wave number k,

$$c_g = \frac{gT}{4\pi}\tanh kh\left(1 + \frac{2kh}{\sinh 2kh}\right),\tag{10}$$

while the positive free wave travels faster, following the shallow water wave velocity ($\sqrt{gh}$). Therefore, the two sets of waves slowly separate. This conceptual model is designed by adopting the definition of the forced wave proposed by Mei [2] (right panel of Figure 1). The definition by Longuet-Higgins and Stewart [1] would not necessarily require a positive free wave to balance the increasing magnitude of the forced wave depression. The results show that for groups with the same mean primary wave frequency (i.e., the same group velocity), the gain (change in the crest to trough height) and phase (lag of the long wave trough relative to the forcing) of the total infragravity wave varies with the group frequency.

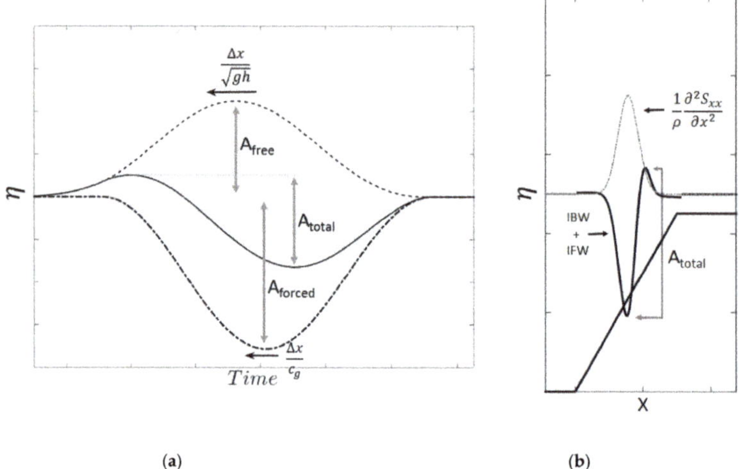

(a)

(b)

Figure 3. (**a**) Schematic representation of the surface elevation at a fixed position for a superposition of forced and free waves. Dot-dashed line - bound wave propagating with group velocity and amplitude A_{forced}. Dashed line - free wave generated during the shoaling process propagating with with respective amplitude A^+_{free}. Full line - total signal with amplitude A_{total}. (**b**) Spatial representation of bound wave shoaling. Black line, superposition of incident bound wave (IBW) and an incident forward free wave (IFW) generated to balance changes in the forced solution during shoaling. Dotted-dashed line is the representation of the radiation stress forcing.

Thus, even though the forced bound wave shoaling is the same as the equilibrium solution, an apparent frequency dependence appears in the shoaling rate of the total long wave. This is illustrated in Figure 4, which shows how frequency influences the total infragravity wave amplitude at a fixed position on the virtual slope. The results are qualitatively similar to the numerical results in Madsen et al. [23], their Figure 20, obtained from a Boussinesq wave model with improved dispersion in deeper water.

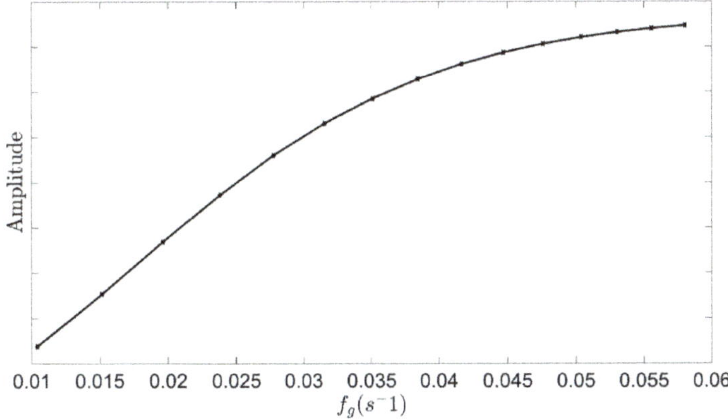

Figure 4. Variation in total infragravity wave amplitude at a fixed position on the virtual slope for wave groups of different frequency using the conceptual model of bound and free wave superposition illustrated in Figure 3. Wave groups have the same group velocity but varying frequency.

In this model approach, the only factor affecting the apparent changes in the shoaling rate of the total infragravity wave is the behaviour of the free wave relative to the wave group (forced wave, Equation (9). Over the same distance, the free waves generated by shorter wave groups will tend to propagate further apart from the bound wave than the free waves generated by longer wave groups, since the difference in phase speed between the forced wave and the free long wave is greater. In addition, for a longer group, the free wave needs to travel a longer distance to get ahead of the bound wave. Note that due to the difference between the free wave speed $\sqrt{gh}$ and group velocity, c_g, a lag is expected for the total long wave (Figure 3a) and the magnitude varies according to the length of the forcing. Interestingly, the lags are also only observed in the time domain, not in the space domain. This is consistent with the observations of Baldock [11] and also with the data in Figure 2.

Based on these lags, hypothetical rates of energy transfer are calculated using Equation (8). As illustrated in Figure 5, the apparent rate of energy transfer also seems to increase with increasing frequency of the forced wave. This is the same behaviour as that predicted by the energy transfer model of Janssen et al. [12]. The apparent shoaling rate for the long wave groups are closer to $\sim h^{-1/4}$, i.e., similar to free long waves, while short groups are closer to $\sim h^{-5/2}$, qualitatively matching the different shoaling behavior in the mild-slope and the steep-slope regimes described by Battjes et al. [13]. The results presented for this conceptual model are quite simple, but they indicate that adding free waves into the bound wave shoaling process may change the results and therefore the interpretation of energy transfers, and phase lags of the forced wave.

In order to verify this hypothesis a set of simulations using the COULWAVE model [35] which adopts the fully nonlinear Boussinesq equations, is presented. This model is widely used for wave modelling and has been well-verified in previous studies, including surf beat studies [36]. The model solves the conservative form of weakly dispersive and fully nonlinear Boussinesq equations in 1-D, following [37]

$$\frac{\partial H}{\partial t} + \frac{\partial H U_\alpha}{\partial x} + D^c = 0 \tag{11}$$

$$\frac{\partial HU_\alpha}{\partial t} + \frac{\partial HU_\alpha^2}{\partial} + gH\frac{\partial \eta}{\partial x} + gHD^x + U_\alpha D^c - R_b^x = 0 \tag{12}$$

where $H = \eta + h$ is the total water depth, η is the surface elevation, h water depth, U_α velocity at water depth z_α. D^c and D^x are the higher order terms that include bottom turbulence and dispersive properties and R_b^x is the wave breaking dissipation term. The simulations are performed introducing the radiation stress forcing term into the momentum Equation (12), using the right hand term in Equation (1), as the driving mechanism of bound wave generation, with radiation of free waves (4) as required to ensure continuity. Following Nielsen and Baldock [8], seven cases of Gaussian-shaped wave group envelopes, similar to Equation (8), with the same primary mean wave period of $T = 6\,s$ (and the same group velocity, Equation (9)), but different group lengths, varying from 200 m to 345 m, are tested. The domain starts with a long section of horizontal bottom ($h = 20\,m$) connected to a slope ($\beta = 0.025$) followed by a long plateau ($h = 5\,m$), similar to Figure 3b. The length of the plateau is sufficient for the free waves, generated during the shoaling process, to propagate completely away from the forced waves so that the different waves are separated in space and time. The initial condition is flat water, with the forced wave at the initial time-step balanced by the two free waves as proposed by Nielsen and Baldock [8]. The length of the initial horizontal domain is also sufficient so the free waves propagate completely away from the forcing region, such that when the forcing zone propagates to the start of the sloping beach, only the forced wave remains as a pure depression, representing the conditions proposed in the model of Mei [2] (Figure 1b).

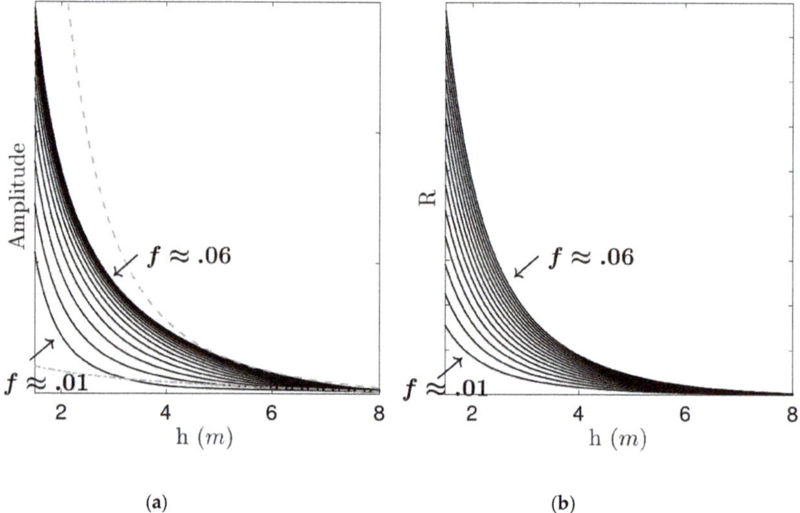

Figure 5. (a) Amplitude variation of total infragravity wave during shoaling for different group frequencies, grey dashed line ($\sim h^{-5/2}$), grey dash-dotted line ($\sim h^{-1/4}$). (b) Variation in theoretical rates of energy transfer calculated based on the observed lags using Equation (7) for different group frequencies.

Similar to the results from the conceptual model, Figure 5, these specific examples, Figure 6, show that in the shoaling zone the shorter infragravity waves have larger apparent amplitudes than the longer infragravity waves. Also, the shorter the long wave, the greater is the apparent lag with respect to the forcing. However, after the forcing reaches the constant depth plateau, the shoaling stops and the free waves propagate away ahead of the forcing region, leaving behind only forced waves, which all have with same amplitude. The differences in amplitude of the leading free waves at around $t \approx 800\,s$ on the plateau, Figure 6b, are due to the different lengths of the forcing relative to the length of the slope where the changes are occurring, with shorter groups resulting in a smaller free wave amplitude.

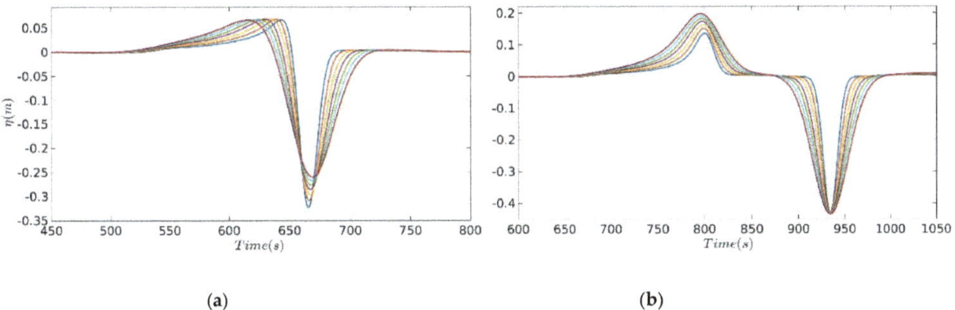

(a)   (b)

Figure 6. Waves generated by Gaussian-shaped wave group envelopes with different forcing length but with the same propagation speed. Each line represents a different group length, from shortest 200 m (blue line), to the longest 345 m (dark red). Shoaling zone (**a**) and horizontal plateau (**b**).

The numerical calculations for these same specific examples also show that for all cases no lag occurs in the spatial domain between the forcing and the bound wave (see Figure 7) but that a lag is observed in the time domain (Figure 8), similar to that observed in the laboratory data (Figure 3). The difference between observations in the spatial domain and the time domain is important, since most measurements of the bound wave lag occur from data obtained in the time-domain. In this model, the forced wave part of the total wave is by definition in antiphase with the forcing, so it does not lag the forcing. However, since the system is unsteady and evolving as the wave passes a fixed location, the total solution has an apparent lag in the time domain. Overall, the results suggests that when considering the approach of Mei [2] (Figure 1b) for the bound wave problem via radiation stress forcing, both the shape and shoaling rates could be affected by the generation of positive leading free waves required to balance the shoaling forced wave. Further investigation is needed in order to verify if free waves generation occurs in laboratory or natural conditions. This is difficult, since long spatial domains are required to observe separation of free and forced waves with very similar celerity. Indeed, in shallow water the waves do not separate at all.

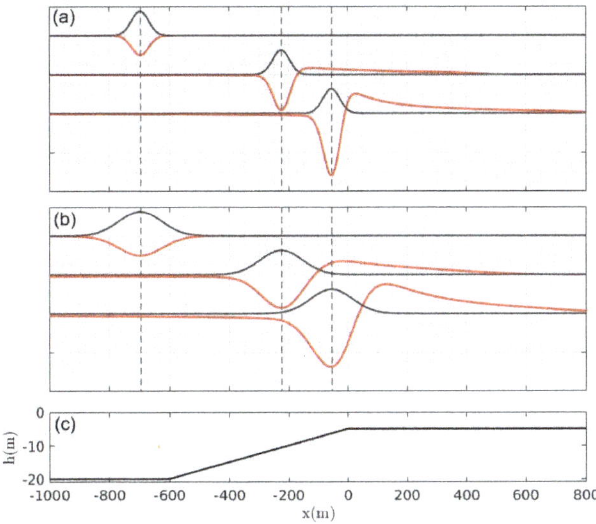

Figure 7. Spatial visualization of a short (**a**) and long (**b**) Gaussian-shape radiation stress (black line) and total long wave surface elevation (red) at different time instants and location over the bottom profile (**c**). The vertical dashed lines indicate the zero lag between forcing and bound wave.

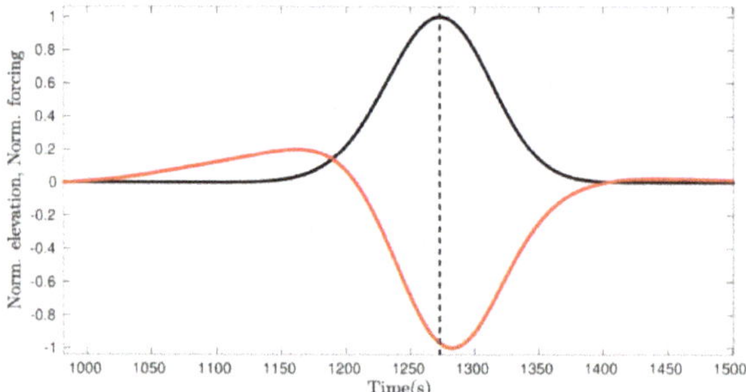

Figure 8. Time series of surface displacement (at x = −100 m in Figure 7c), normalized by its absolute minimum value (red) and Gaussian-shape radiation stress normalized by its maximum value (black), for the longer forcing case shown in Figure 7b.

3. Bound Wave Dissipation after Short Wave Breaking

One important question yet to be fully answered is what happens to the bound wave after short wave breaking? Different datasets indicate bound wave release [12,13], while others indicate dissipation of the forced wave in the surf zone [17,24,36,38]. Baldock [15] suggested that the relationship between short and infragravity wave with water depth at the breakpoint is important to determine whether the bound wave is released or if it remains forced, decaying with the dissipation of the short waves. During the breaking process, three different conditions are possible: (i) at the breaking point both the short and forced waves are not in shallow water, $c_g < \sqrt{gh}$ and the length of the bound wave is not much longer than h; (ii) the bound wave is much longer than h, but the short waves are not, hence the bound wave number differs from the wave number of a free wave with same frequency; (iii) both short and bound wave are in shallow water at the break point. Bound wave release is considered to occur when both short and bound waves are in shallow waters, therefore being strictly valid only for the third condition. While the first possibility is more likely for deep-water wave breaking, the second condition can occur at the edge of the surf zone, especially during storm conditions [15].

Baldock and O'Hare [39] compared the behaviour of free non-breaking waves and forced infragravity waves inside the surf zone. The laboratory experiment was designed so the effects of short wave breaking on forced and free waves could be evaluated. With one set of cases based on two free waves, a short breaking wave and a smaller and longer non-breaking wave; and a second set of cases based on fully modulated bichromatic groups. The results have shown that while small or nearly no dissipation occurred for the free non-breaking waves, strong dissipation occurred for short bound waves starting at the break point. The authors then observed that the short wave breaking was occurring when the waves were not in shallow water, justifying the forced wave decay in the surf zone. The numerical investigation presented in Moura and Baldock [36] showed, for the same cases, that no significant breakpoint excursion occurs at the group frequencies, therefore there was no breakpoint forcing at these frequencies. This removes the possibility of interference between the bound wave and breakpoint force long waves, which has been previously suggested as an explanation for the infragravity wave amplitude decay in the surf zone [3].

In order to test whether the breaking depth affects the behaviour of the bound wave, the simulations using the COULWAVE model for two fully-modulated bichromatic cases presented by Moura and Baldock [36] are extended to a larger range of wave conditions, where the amplitude of the fully-modulated wave groups are progressively reduced, hence shifting the break point to shallower water. The COULWAVE model has two options to simulate wave breaking, the eddy viscosity model described in Kennedy et al. [40] and an adaptation to this model, named the transport-based model [41], which is the one selected for this study. Both models use the same eddy viscosity dissipation term and a triggering mechanism (local time derivative of the surface elevation exceeding a threshold) to initiate wave breaking. For further details of the modelling see Moura [42]. For both cases, strong bound wave dissipation, starting at the edge of the surf zone, have been observed in the data [24] and reproduced numerically [36]. The numerical results in Figure 9 indicate that independently of the water depth at the short wave breakpoint the bound wave is dissipated from the breakpoint up to the swash zone. The small outgoing waves corroborate that the bound wave dissipation occurs for all tested cases. This means that when short wave breaking occurs in shallow water, and hence bound wave release is expected, strong dissipation is still occurring. This implies that the energy dissipation is due to turbulence generated by the short waves. This supports the suggestion of Battjes et al. [13], who proposed that infragravity waves that are relatively short compared to the surf zone width are more easily dissipated than longer waves whose its lengths are only a fraction of the surf zone. Hence, dissipation is also strongly frequency dependent.

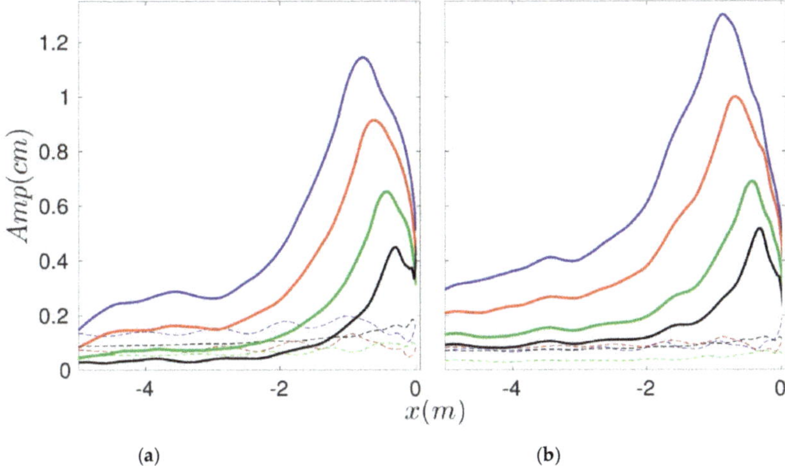

Figure 9. Incident (full lines) and outgoing (dash lines) infragravity surface elevation at f_g for bichromatic cases 1055 (**a**), 1060 (**b**) from Baldock et al. [24]. Colours are for different fully modulated wave groups with short wave amplitude 0.05 m (blue), 0.035 m (red), 0.025 m (green) 0.015 m (black).

To extend this analysis further, using the COULWAVE model, and following the experiment layout in Baldock and O'Hare [39], the effects of a surf zone on the dissipation of non-breaking free long waves is numerically investigated by two sets of simulations, with and without a surf zone. The first set of simulations with a surf zone comprise of a short breaking monochromatic wave ($w1$) with an amplitude of 5 cm and frequency of 0.8 Hz paired with longer free monochromatic waves ($w2$) with an amplitude of 0.05 cm and frequencies varying from 0.09 to 0.5 Hz, to represent free long waves inside a surf zone. The second set of simulations comprised of the same monochromatic free long waves ($w2$) but with no short waves and hence no surf zone. The model results for the long waves ($w2$) with and without a surf zone are then compared in terms of total, incident and outgoing wave amplitude.

The standing wave total amplitude patterns generated by $w2$ in the absence or presence of a surf zone are significantly different (Figure 10). While for both scenario a reduction of the nodal structures occurs towards the higher frequencies, the process is intensified for the cases with a surf zone present. Thus, dissipation of free long waves by eddy turbulence inside a surf zone is certainly frequency dependent based on the COULWAVE model, and it is likely this explains the decay of the incident forced wave also, as proposed by Battjes et al. [13], with some additional reduction in amplitude since the forcing is also being reduced by short wave breaking.

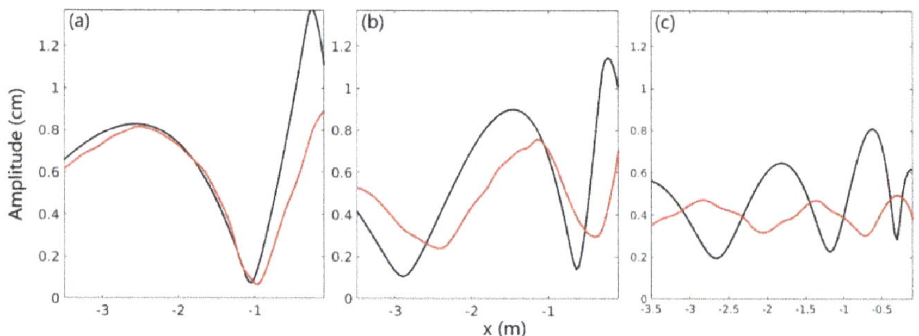

Figure 10. Cross-shore nodal pattern for long waves of different frequencies with (red) and without (black) a surf zone. Frequencies vary from f = 0.195 (**a**), 0.265 (**b**) and 0.44 (**c**) Hz.

4. Discussion

There are two interpretations of the forced wave in the literature, as indicated in Figure 1. The model proposed by Mei [2] appears to require the radiation of free waves if the forced wave is to change amplitude by shoaling, which is suggested by Nagase and Mizuguchi [21] and Nielsen and Baldock [8]. However, radiation of such free waves during shoaling has yet to be observed with any degree of certainty. The model of Longuet-Higgins and Stewart [1] does not appear to require such radiation of free waves to ensure continuity, since the surface elevation has bot negative and positive components, both of which can grow. This study has adopted the conceptual model approach of Mizuguchi and Nagase and Nielsen and Baldock and constructed a simple analytical model to investigate the shoaling behavior of the total long wave if forced and radiating free waves co-exist. The results clearly show an apparent growth of the long wave at rates less than that proposed by Longuet-Higgins and Stewart (Figure 5), which has been observed in many data sets, and theoretically analysed by alternate means by Janssen et al. [12] and Battjes et al. [13] The growth rate and lag of the long wave behind the forcing are frequency dependent, in agreement with the mild and steep slope shoaling regimes proposed by Battjes et al. However, this analysis shows that the lag only occurs in the time domain, which is how most data is collected, but not in the spatial domain. The same qualitative results were found when performing simulations with the COULWAVE model using the radiation stress term in the momentum equation to force the generation and propagation of bound waves. This is again supported by the limited data sets available, see Figure 3. With this model approach, the forced waves (not the total wave) follow the equilibrium solution of Longuet-Higgins and Stewart, and the shoaling of the forced wave part is not frequency dependent. We note that radiation of free waves during shoaling has yet to be verified in observations, and further work with RANS models is ongoing to determine if this is the case.

We have also used the COULWAVE to extend experimental data sets that investigated the dissipation of the incident forced and free long waves in the surf zone. The results clearly suggest that the dissipation rate is dependent on the long wave frequency or length, for both forced and free waves, in agreement with Battjes et al. [13] irrespective of the depth at which short wave break. This suggests that the reduction in forced wave amplitude inside the surf zone is a combination of the reduced forcing, as originally proposed by Longuet-Higgins and Stewart [1] and further analysed by Baldock [15], and turbulent dissipation as suggested by Battjes et al [13].

Author Contributions: T.M. and T.E.B. conceived and designed the modelling; T.M. performed the modelling and analysed the results; T.M. and T.E.B. wrote the paper.

Funding: Theo Moura gratefully acknowledges a scholarship from the Brazilian National Council for Scientific and Technological Development (CNPq). This work was supported by Australian Research Council grants LP100100375 and DP13101122.

Conflicts of Interest: The authors declare no conflict of interest.

References

1. Longuet-Higgins, M.S.; Stewart, R.W. Radiation stress and mass transport in gravity waves, with application to 'surf beats'. *J. Fluid Mech.* **1962**, *13*, 481–504. [CrossRef]
2. Mei, C.C. *The Applied Dynamics of Ocean Surface Waves*; World Scientific: Singapore, 1989; p. 740.
3. Schäffer, H.A. Infragravity waves induced by short-wave groups. *J. Fluid Mech.* **1993**, *247*, 551. [CrossRef]
4. Proudman, J. The Effects on the Sea of Changes in Atmospheric Pressure. *Geophys. J. Int.* **1929**, *2*, 197–209. [CrossRef]
5. Whitham, G.B. Mass, momentum and energy flux in water waves. *J. Fluid Mech.* **1962**, *12*, 135–147. [CrossRef]
6. Molin, B. On the generation of long-period second-order free-waves due to changes in the bottom profile. *Ship Res. Inst. Pap.* **1982**, 1–23.
7. Mei, C.C.; Benmoussa, C. Long waves induced by short-wave groups over an uneven bottom. *J. Fluid Mech.* **1984**, *139*, 219. [CrossRef]
8. Nielsen, P.; Baldock, T.E. И-Shaped surf beat understood in terms of transient forced long waves. *Coast. Eng.* **2010**, *57*, 71–73. [CrossRef]
9. Tinti, S.; Bortolucci, E.; Chiavettieri, C. Tsunami Excitation by Submarine Slides in Shallow-water Approximation. *Pure Appl. Geophys.* **2001**, *158*, 759–797. [CrossRef]
10. Didenkulova, I.; Nikolkina, I.; Pelinovsky, E.; Zahibo, N. Tsunami waves generated by submarine landslides of variable volume: Analytical solutions for a basin of variable depth. *Nat. Hazards Earth Syst. Sci.* **2010**, *10*, 2407–2419. [CrossRef]
11. Baldock, T.E. Long wave generation by the shoaling and breaking of transient wave groups on a beach. *Proc. R. Soc. A Math. Phys. Eng. Sci.* **2006**, *462*, 1853–1876. [CrossRef]
12. Van Dongeren, A.R.; Janssen, T.T.; Battjes, J.A. Long waves induced by short-wave groups over a sloping bottom. *J. Geophys. Res. Space Phys.* **2003**, *108*, 108.
13. Battjes, J.A.; Bakkenes, H.J.; Janssen, T.T.; Van Dongeren, A.R. Shoaling of subharmonic gravity waves. *J. Geophys. Res. Space Phys.* **2004**, *109*, 109. [CrossRef]
14. Dong, G.; Ma, X.; Perlin, M.; Ma, Y.; Yu, B.; Wang, G. Experimental study of long wave generation on sloping bottoms. *Coast. Eng.* **2009**, *56*, 82–89. [CrossRef]
15. Baldock, T.E. Dissipation of incident forced long waves in the surf zone—Implications for the concept of "bound" wave release at short wave breaking. *Coast. Eng.* **2012**, *60*, 276–285. [CrossRef]
16. Padilla, E.M.; Alsina, J.M. Transfer and dissipation of energy during wave group propagation on a gentle beach slope. *J. Geophys. Res. Oceans* **2017**, *122*, 6773–6794. [CrossRef]
17. Padilla, E.M.; Alsina, J.M. Long Wave Generation Induced by Differences in the Wave-Group Structure. *J. Geophys. Res. Oceans* **2018**, *123*, 8921–8940. [CrossRef]
18. Van Dongeren, A.R.J.A.; Battjes, J.; Janssen, T.; Van Noorloos, J.; Steenhauer, K.; Steenbergen, G.; Reniers, A.J.H.M. Shoaling and shoreline dissipation of low-frequency waves. *J. Geophys. Res.-Oceans* **2007**, *112*, C02011. [CrossRef]
19. Bowen, A.J.; Henderson, S.M. Observations of surf beat forcing and dissipation. *J. Geophys. Res. Space Phys.* **2002**, *107*, 107.
20. Raubenheimer, B.; Thomson, J.; Elgar, S.; Herbers, T.H.C.; Guza, R.T. Tidal modulation of infragravity waves via nonlinear energy losses in the surfzone. *Geophys. Res. Lett.* **2006**, *33*. [CrossRef]
21. Nagase, S.; Mizuguchi, M. Laboratory experiment on long wave generation by time-varying breakpoint. *Coast. Eng.* **1997**, 1307–1320.
22. Masselink, G. Group bound long waves as a source of infragravity energy in the surf zone. *Cont. Shelf Res.* **1995**, *15*, 1525–1547. [CrossRef]
23. Madsen, P.A.; Sørensen, O.; Schäffer, H. Surf zone dynamics simulated by a Boussinesq type model. Part II: Surf beat and swash oscillations for wave groups and irregular waves. *Coast. Eng.* **1997**, *32*, 289–319. [CrossRef]
24. Baldock, T.E.; Huntley, D.A.; Bird, P.A.D.; O'hare, T.; Bullock, G.N. Breakpoint generated surf beat induced by bichromatic wave groups. *Coast. Eng.* **2000**, *39*, 213–242. [CrossRef]
25. Baldock, T.E.; Huntley, D.A. Long-wave forcing by the breaking of random gravity waves on a beach. *Proc. R. Soc. Lond. A. Phys. Eng. Sci.* **2002**, *458*, 2177–2201. [CrossRef]

26. Orszaghova, J.; Taylor, P.H.; Borthwick, A.G.; Raby, A.C. Importance of second-order wave generation for focused wave group run-up and overtopping. *Coast. Eng.* **2014**, *94*, 63–79. [CrossRef]
27. Mansard, E.; Barthel, V. Shoaling properties of bounded long waves. *Coast. Eng.* **1984**, *1985*, 798–814.
28. List, J.H. A model for the generation of two dimensional surf beat. *J. Geophys. Res.* **1992**, *97*, 5623–5635. [CrossRef]
29. Longuet-Higgins, M.S.; Stewart, R. Changes in the form of short gravity waves on long waves and tidal currents. *J. Fluid Mech.* **1960**, *8*, 565–583. [CrossRef]
30. Phillips, O.M. *The Dynamics of the Upper Ocean*; CUP Archive, 1966.
31. Henderson, S.M.; Guza, R.T.; Elgar, S.; Herbers, T.H.C. Refraction of Surface Gravity Waves by Shear Waves. *J. Phys. Oceanogr.* **2006**, *36*, 629–635. [CrossRef]
32. Guedes, R.M.; Bryan, K.R.; Coco, G.; Bryan, R.; Coco, G. Observations of wave energy fluxes and swash motions on a low-sloping, dissipative beach. *J. Geophys. Res. Oceans.* **2013**, *118*, 3651–3669. [CrossRef]
33. Bowers, E.C. Low Frequency Waves in Intermediate Water Depths. In Proceedings of the 23rd International Conference on Coastal Engineering, Venice, Italy, 4–9 October 1992; pp. 832–845.
34. Van Leeuwen, P.J. Low Frequency Wave Generation Due to Breaking Wind Waves. Ph.D. Thesis, Delft University of Technol, Delft, The Netherlands, 1992.
35. Lynett, P.J. Nearshore wave modeling with high-order Boussinesq-type equations. *J. Waterw. Port Coast. Ocean Eng.* **2006**, *132*, 348–357. [CrossRef]
36. Moura, T.; Baldock, T.E. New evidence of breakpoint forced long waves: Laboratory, numerical, and field observations. *J. Geophys. Res. Oceans.* **2018**, *123*, 2716–2730. [CrossRef]
37. Kim, D.-H.; Lynett, P.J.; Socolofsky, S.A. A depth-integrated model for weakly dispersive, turbulent, and rotational fluid flows. *Ocean Model.* **2009**, *27*, 198–214. [CrossRef]
38. Contardo, S.; Symonds, G. Infragravity response to variable wave forcing in the nearshore. *J. Geophys. Res. Oceans* **2013**, *118*, 7095–7106. [CrossRef]
39. Smith, J.M.; Baldock, T.E.; O'Hare, T.J. Energy transfer and dissipation during surf beat conditions. *Coast. Eng.* **2005**, *4*, 1212–1224.
40. Kennedy, A.B.; Chen, Q.; Kirby, J.T.; Dalrymple, R.A. Boussinesq Modeling of Wave Transformation, Breaking, and Runup. I: 1D. *J. Waterw. Port Coast. Ocean Eng.* **2000**, *126*, 39–47. [CrossRef]
41. Løvholt, F.; Lynett, P.; Pedersen, G. Simulating run-up on steep slopes with operational Boussinesq models; capabilities, spurious effects and instabilities. *Nonlinear Process. Geophys.* **2013**, *20*, 379–395. [CrossRef]
42. Moura, T. *Infragravity Wave Forcing in the Surf and Swash Zone*; School of Civil Engineering, the University of Queensland: Brisbane, Australia, 2016.

Journal of
*Marine Science
and Engineering*

Article

Downscaling Future Longshore Sediment Transport in South Eastern Australia

Julian O'Grady [1,*], Alexander Babanin [2] and Kathleen McInnes [1]

[1] Commonwealth Scientific and Industrial Research Organisation (CSIRO), Oceans and Atmosphere, Melbourne, Victoria 3195, Australia
[2] Department of Infrastructure Engineering, The University of Melbourne, Melbourne, Victoria 3052, Australia
* Correspondence: julian.ogrady@csiro.au

Received: 29 June 2019; Accepted: 20 August 2019; Published: 26 August 2019

Abstract: Modelling investigations into the local changes in the shoreline resulting from enhanced atmospheric greenhouse gas concentrations and global climate change are important for supporting the planning of coastal mitigation measures. Analysis of Global Climate Model (GCM) and Regional Climate Model (RCM) simulations has shown that Lakes Entrance, a township located at the northern end of Ninety Mile Beach in south-eastern Australia, is situated in a region that may experience noticeable future changes in longshore winds, waves and coastal currents, which could alter the supply of sediments to the shoreline. This paper will demonstrate a downscaling procedure for using the data from GCM and RCM simulations to force a local climate model (LCM) at the beach scale to simulate additional nearshore wind-wave, hydrodynamic and sediment transport processes to estimate future changes. Two types of sediment transport models were used in this study, the simple empirical coastline-type model (CERC equation), and a detailed numerical coastal area-type model (TELEMAC). The two models resolved transport in very different ways, but nevertheless came to similar conclusions on the annual net longshore sediment transport rate. The TELEMAC model, with the Soulsby-Van Rijn formulation, showed the importance of the contribution of storm events to transport. The CERC equation estimates more transport during the period between storms than TELEMAC. The TELEMAC modelled waves, hydrodynamics and bed-evolutions are shown to agree well with the available observations. A new method is introduced to downscale GCM longshore sediment transport projections using wave-transport-directional change parameter to modify directional wave spectra. We developed a semi-empirical equation (NMB-LM) to extrapolate the ~3.7-year TELEMAC, storm dominated transport estimates, to the longer ~30-year hindcast climate. It shows that the shorter TELEMAC modelled period had twice as large annual net longshore sediment transport of the ~30 year hindcast. The CERC equation does not pick up this difference for the two climate periods. Modelled changes to the wave transport are shown to be an order of magnitude larger than changes from storm-tide current and mean sea level changes (0.1 to 0.2 m). Discussion is provided on the limitations of the models and how the projected changes could indicate sediment transport changes in the nearshore zone, which could impact the coastline position.

Keywords: longshore transport; climate change; downscaling

---

1. Introduction

Investigating and predicting changes in the shoreline is important for supporting the planning of coastal mitigation measures for public and private infrastructure from severe storms—particularly under the influence of anthropogenic climate change [1].

When waves approach the coastline at an oblique angle, they dissipate in the shallowing water and create a force in the direction parallel to the shoreline, which—if strong enough—can result in longshore sediment transport. Longshore sediment transport is a combination of; (1) longshore currents driven by these depth-induced wave forces, as well as tidal and wind driven flow and; (2) beach drift from the swash action of individual waves running up and falling down the coastline at an angle in a zig zag pattern.

Sea level rise has been considered the defining driver of coastline change with long-term climate change since the inception of the simple assumption of maintaining the cross-shore profile [2]. While important, some recent regional studies have shown longshore transport to be more important to barrier erosion than sea level rise, e.g., for the Danish North Sea coast [3], and on the Florida coast in the United States of America, where [4] found that coastlines have advanced (accreted) instead of retreated at a rate attributable to recorded sea level rise.

The study site, Ninety Mile Beach (NMB) is a 144 km long sandy barrier dune system that is located in south-eastern Australia (Figure 1). At the eastern end of the beach is a man-made ocean entrance to the Gippsland Lakes located at the township of Lakes Entrance. Previous studies have shown that Lakes Entrance is situated in a region that may experience noticeable future changes in longshore winds, waves and currents, which could alter the supply of sediments to the shoreline [5]. The study site has significant financial and environmental value [6].

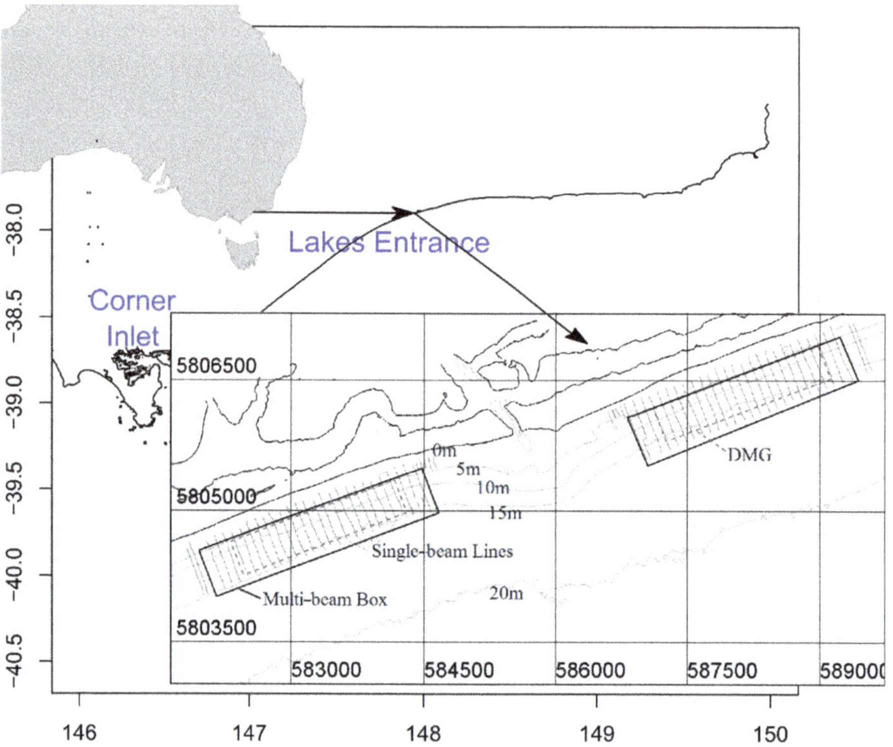

Figure 1. Map of Ninety Mile Beach in southeastern Australia (nautical coordinates) with an overlay of Australia (top right) and an insert of bathymetric surveys at Lakes Entrance (bottom right, WGS84 zone 55 coordinates). The insert of Lakes Entrance provides examples of the single-beam transects (cross-shore lines separated by 100 m) and multi-beam survey area (thick black box) at both the west and east dredge material grounds (DMG) sites (grey dashed box).

The goal of this study is to better resolve the present-day longshore transport climate by setting-up sediment transport models of different complexity. Nearshore wave and hydrodynamic processes are not resolved in Global Climate Model (GCM) simulations, so downscaling methods are required to conduct climate change analysis near the shoreline (e.g., [7,8]). We first validate a detailed numerical model which we then use with empirical models to investigate the downscaled longshore transport climate of a high impact future greenhouse gas (GHG) scenario. This is done to understand the contribution of waves, currents and sea level rise to future changes to the coast. In the flowing section (Section 2) a review of sediment transport modelling is provided, followed by a review of downscaling methods. Sections 3 and 4 describe the data and modelling methodology applied for our study for NMB and Sections 5 and 6 provide the results, discussion and conclusions of the study.

2. Methodological Review

2.1. Longshore Sediment Transport Models

One line empirical equations—which are derived from regression analysis of measured sediment transport—have been used to find the simplest relationship (or pattern of behaviour) between sediment transport and wave and current parameters (e.g., [9–14]. The limitation for some empirical equations, for example, is that they have been calibrated to locations where transport is almost exclusively controlled by waves and therefore do not account for coastal (tide and storm) current parameters (e.g., [15]). These equations may underestimate the contribution of wind- and tide- driven currents when applied to regions of strong tide and storm surge currents [11].

Coastal numerical sediment transport models including fluid dynamic processes (e.g., Navier Stokes equations), were originally developed for river flows. Since the 1980s, these models have been applied to the coastal ocean domain with the important inclusion of surface gravity wind-wave effects [16]. These coastal sediment transport models, typically implemented in sizable software packages, employ computational fluid dynamics, through numerical computations of partial differential equations (PDEs) on a horizontal grid, to solve both the longshore and cross-shore dynamics (e.g., ADCIRC, ROMS, MIKE-21, XBEACH and TELEMAC see model comparison in [17]). The main advantages of these coastal area models are that the hydrodynamic flow and waves are more accurately resolved, and that the sediment budget can be balanced over the entire coastal area. Coupled models can take on the challenge of the most complex hydrodynamic processes in the nearshore region, in particular, the effect of depth-induced wave breaking. When ocean waves enter the nearshore region, they feel the shallowing depths and steepen as they slow down—until they eventually break. Along with generating turbulence, the effect of this breaking results in a stress applied to the water column, causing rips- and longshore-currents which scatter sediment particles [18,19]. Superimposed on the mean coastal current circulation field is the orbital motion of waves—which can mobilise sediments. Wave orbital motion occurs at the sub-grid-scale for practical model applications, and is typically formulated from linear wave theory [20]. The modelling of sediment transport is dependent on the flow or stress (quadratic flow/stress law) exceeding a critical threshold to mobilise sediments. For this reason, the coincidence of strong wave orbital flow with wind-, wave- and tide-driven currents above a critical flow threshold will result in sediment transport. For practical application of numerical models, sediment transport typically occurs at the sub-grid scale. Therefore, they require parameterisation from empirical equations, similar to all types of sediment transport models with varying complexity. Other models with reduced numerical complexity, i.e., models that do not simulate the detailed PDEs such as the Navier-Stokes equations, have shown to replicate long term shoreline change [21–23].

The choice of model complexity depends on the scale of study and what processes need to be resolved, from beach change due to individual storm event to the evolution of a coastline including climate change impacts. The choice of model also depends on the complexity of the coastline, from a simple uniform sandy beach, to a coastline with longshore hard structures (breakwaters and groins), to tidal basins, mega-renourishment and land reclamations. Careful consideration must be made regarding the type of model to use and the value of resources allocated to running the model so as to make efficient predictions or to study detailed processes.

In this study, the CERC energy flux equation is investigated as it has been widely used for this location—including the dredging program at the field site [24] and a previous study on GCM-derived transport changes [5]. The TELEMAC coastal area type model is used in this study for personal preferences, as it is open source; has unstructured grid configuration; all model packages (wave-hydrodynamic-sediment) have two-way coupling and it has high quality code formatting conventions across all files within the system and detailed model documentation with an active user community forum. The TELEMAC suite of models has been shown to provide similar results to the proprietary Mike-21 model for the Italian coast [25]. An example of TELEMAC's ability to simulate multiyear, decadal and longer timescales has been studied for the German coastline [26]. Climate change downscaling of RCMs with TELEMAC has been conducted for the English coast [27]. These regional studies demonstrate the variety of approaches, high degree of empiricism and challenges of modelling long-term sediment transport with limited computational resources.

2.2. Downscaling

Investigation of future projected GCM impacts at the local scale requires some method of modifying and refining the large-scale gridded GCM predictions down to the small-scale—a process known as downscaling [28,29]. The downscaling of wind, waves and hydrodynamics has been previously reviewed by [30]. Figure 2 provides a flow chart of different approaches to downscaling the RCP radiative forced GCM data down to the local scale, where each method (path through the flow chart) has a different level of complexity, practicality and accuracy. The methods can be categorised into three different approaches described in the next subsections.

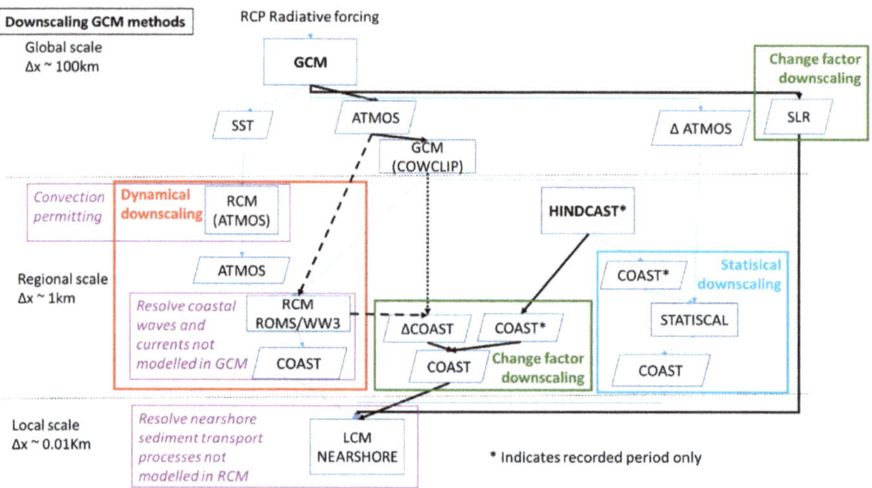

Figure 2. Flow chart of some possible methods for downscaling coastal (COAST) and nearshore (NEARSHORE) wave and current data. Boxed text indicates process (numerical or statistical models) and parallelograms indicated datasets (future and baseline periods). Black arrows indicate the method used in this study, solid for processing both wave and currents, dashed for currents only and dotted for waves only. Blue arrows indicate other possible downscaling methods not used in this study.

2.2.1. Dynamical Downscaling

Dynamical downscaling utilises a direct time series of gridded GCM output, usually bias-adjusted sea surface temperature (SST), to force a high resolution nested atmospheric circulation RCM (e.g., [31]). High resolution atmospheric RCMs can dynamically simulate internal processes such as atmospheric convection—which is otherwise parametrised in course resolution GCMs [32]. High resolution surface winds from atmospheric RCMs have been used to force regional wave RCMs [7,33] and regional hydrodynamic RCMs [34]. Dynamic downscaling has also been used to describe the one-way-coupling of wave models run at the global scale of GCMs wind fields [35] to other wave models run at a higher resolution to resolve processes not simulated in the global model for coastal regions [36,37]. Statistically significant climate trends in the hydrodynamics have been found without the need for the atmospheric RCM dynamical downscaling step [38]. Models that run without the atmospheric downscaling step have the same limitations of the GCM, notably the inability to resolve the intensities of tightly structured storm systems such as tropical and extratropical cyclones. The dynamic downscaling of sediment transport modelling with a regional atmospheric model has been investigated for data poor and data rich locations and provides examples of where changes to wave-driven longshore transport will have a larger impact then changes to mean sea level [39,40]. Dynamical downscaling can produce an internally consistent model recreation/realisation of historic (baseline) and future climate changes at the expense of large computer resources to compute the simulations and store the datasets. While is it is easy to comprehend that downscaled changes in wave height will have an impact on longshore transport, a review of the effect of wave direction changes on longshore transport projections for European coastlines highlights the relatively large impact that small projected changes in wave direction can have on longshore transport ([37] and references therein). Wave direction was identified as an important driver of erosion for a beach in south-eastern Australia [41]. A study for the coastline of France [37] used multiple GHG GCM simulations, and indicated that the inter-model variability between models with the same GHG pathway could be larger than the variability of a single model with different GHG pathway simulations. Longshore transport projections for the Spanish coastline also provided an example of how the inter-model variability of wave height and direction variables, are accentuated in terms of the single CERC transport variable [42]. Unfortunately, much of the analysis of future wave-climate change is somewhat limited to studies that focus on the parameter of significant wave height [43]. Future mean sea level projections are typically directly applied from global mean GCM output, but can be dynamically downscaled to resolve regional effects [44].

2.2.2. Statistical Downscaling

Statistical downscaling uses a simpler method by taking the observed spatial pattern of variability and then applying statistical techniques to infer local-scale changes from large-scale changes generated by GCMs. The statistical methods include regression models, weather typing and weather generators [29]. For example, waves have been statistically downscaled with a regression method for extreme value distributions [45] and hydrodynamics have been statistically downscaled with a regression model based on canonical correlation analysis [46]. The Lookup tables, clustering and hybrid methods that have been used to generate wave conditions for shoreline change can be considered to be aligned with statistical downscaling methods of weather typing and weather generators [21,47].

2.2.3. Change Factor (CF) Downscaling

The Change Factor (CF) downscaling method is the simplest method [48]. It takes a time series of the observed (measured or hindcast) climate and applies an offset based on the difference between the model's future and baseline mean periods. The CF method is usually a single value of change applied to the observed time series, but could also be applied as a seasonal or monthly mean change value.

Each of the downscaling methods has their upsides and limitations [29]. The CF method was chosen for our study of NMB because it allowed us to build on the previous analysis on the large scale changes [5] and because of the availability of regional climate and hindcast simulations. These regional hindcast simulations better capture the extreme weather forcing, where GCMs are known to do a poor job of resolving the intensities of tightly structured storm systems such as East Coast Low storm events, which can bring strong easterly winds to the region in the winter (e.g., [49,50]).

3. Datasets

All datasets and supplementary information are described in detail in the supporting dissertation paper [51]. A general overview is provided here for orientation of the analysis.

3.1. Wave and Hydrodynamic Measurements

Observations of the integrated spectral wave parameters of significant wave height (*Hs*), mean wave period (*Tm*) and mean wave direction (*Dm*) were measured from April 2008 to July 2012 by a directional wave buoy located 2 km off the coast at the Entrance to the Gippsland Lakes in the vicinity of the coastal current measurements in approximately 22 m of water.

Depth-averaged 3D coastal current and water level observations are available from 18 two-month deployments between April 2008 and September 2010. The measurements were from a bottom mounted Teledyne RDI Workhorse Sentinel Acoustic Doppler Current Profiler (ADCP) located at a depth of 21 m, 2 km off the coast at Lakes Entrance. The current meter was located in the vicinity of the wave buoy.

3.2. Regional Wave and Hydrodynamic Model Forcing Data

2D frequency and directional spectral wave data was sourced from the Centre for Australian Weather and Climate Research (CAWCR) WaveWatchIII (WW3) wave hindcast [52]. This dataset was simulated with the WW3 model, v4.08 at 4 arc-minute resolution around the Australian coast.

The hydrodynamic model (ROMS) simulated the depth-averaged tide and atmospheric forced currents and water levels around Australia at a 5 km resolution [53]. Change factor values for both the longshore wave and current transport projections for the HadGEM2-ES, ACCESS1.0, INMCM4 and CNRM-CM5 GCM projections were sourced from the analysis in [5].

3.3. Morphologic Measurements

Three types of bathymetric surveys were used in this study. An airborne LiDAR survey of the entire NMB seaboard was conducted from November 2008 to April 2009 and provided a 2.5 m resolution gridded elevation dataset over land to a depth of approximately 20 m [54]. As part of the ongoing dredging operations at Lakes Entrance, ten bathymetric depth single beam sonar surveys of two large rectangular Dredge Material disposal Grounds (DMGs) and bordering regions of the open coast were undertaken from June 2009 to March 2012 [24]. These 1 m surveys span ~2 km along the coast and extend from ~100 to ~800 m off the coast, framing the design DMG with a 100 m ribbon of extra survey area (Figure 1). One metre Multi-beam soundings superseded the single-beam surveys after 2011. They are available for both sites for five surveys, each between 2012 and 2013, and cover the domains shown in Figure 1.

Field work by [55] sampled marine sediments at Eastern Beach to show constant medium mean grain sizes of around 0.3–0.4 mm from the coastline to 10 m depth. From 10 to 20 m depth the grain size increases somewhat linearly to coarse mean grain sizes of around 0.8 mm at 20 m depth.

4. Methodology

As a general overview, the model simulations were run between the 15 surveys (10 single beam surveys and 5 multibeam). Each of the 14 simulations were initiated with updated survey bathymetry and forced at the ocean boundary by the hindcast datasets to create a ~3.7 years model dataset of the survey period. The simulations were then repeated with a change factor applied to ocean boundary conditions to simulate the future sediment transport climate. In the next three subsections the nearshore model setup is described, followed by the novel method for downscaling GCM derived longshore transport changes to directional wave spectrum and then a new semi-empirical equation to extrapolate the nearshore model to the ~30 year hindcast is presented.

4.1. Nearshore Numerical Local Climate Model

OpenTELEMAC, TELEMAC-MASCARET or just TELEMAC is a suite of finite element numerical programs to solve geophysical fluid dynamics [56]. Since January 2010, the software suite became open source and now has a strong community supporting it. A summary of TELEMAC's ability to model sediment transport is given in [57]. A detailed summary of the model setup used is described in [51]. Three programs from the TELEMAC (version 7.1r1) suite are required to model sediment transport in the nearshore region. The first is TELEMAC2D, used to model the flow hydrodynamics, the second is TOMAWAC, used to model the sea-swell waves and the third is SISYPHE, used to model sediment transport and seabed evolution. Figure 3 shows the design of the model coupling, boundary forcing and downscaling setup.

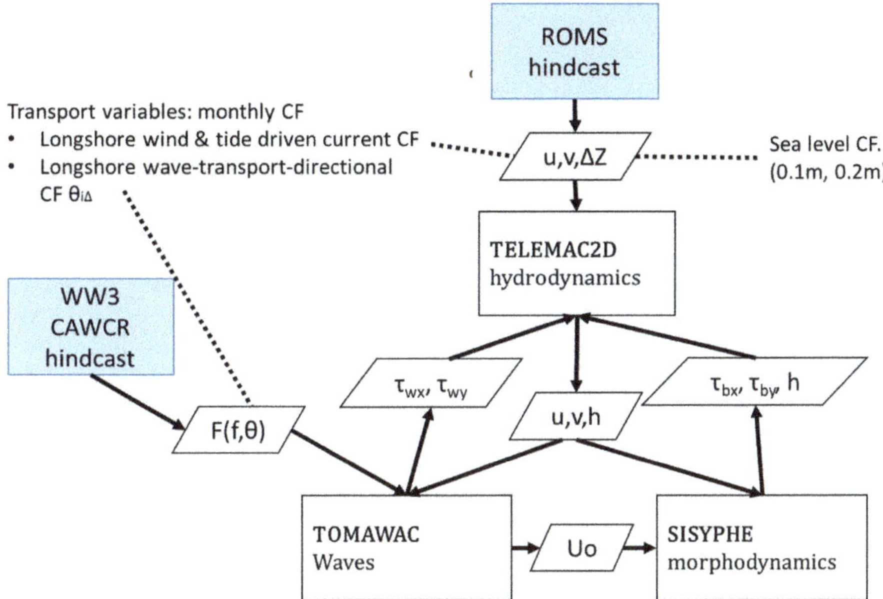

Figure 3. LCM Nearshore model configuration. The flow chart corresponds to the "LCM NEARSHORE" process at the bottom of Figure 2. Equation (1) was used to apply the CF downscaling method to both the ROMS hindcast longshore wind and tide driven currents and to rotate the WW3 2D spectrum by the wave-transport-directional CF. The addition of 0.1 or 0.2 m to the hindcast sea level simulated the effect of sea level rise.

The TELEMAC2D code solves depth-averaged free surface flow Navier-Stokes equations first solved by Saint-Venant [58]. Boundary conditions of water levels and currents were sourced from the 5-km-resolution, regional hindcast of storm tides around Australia using the ROMS hindcast [53]. Longshore coastal currents were generated within the grid by applying a longshore gradient in the water levels from two neighbouring ROMS grid points. Additionally, the flow was modified by prescribing flow velocities in and out of the boundary. In preliminary model setups, the coastal-current velocities and water levels were prescribed at all the boundary points, but yielded poor advection into the domain because the boundary was too constrained. In order to free up the flow constraint, the model source code was modified, to prescribed free flow boundaries depending on the down flow direction of the ROMS flow velocities.

Wave are resolved with TOMAWAC, a third generation spectral wave model [59]. The 2D spectral wave model TOMAWAC was forced at the open sea boundary by 2D (direction and frequency) spectral output from WW3 hourly $\frac{1}{2}$-degree gridded global wave hindcast. TOMAWAC provides four methods for parametrising depth-induced wave breaking. Sensitivity tests in [51] led to the selection of the frequently used equations in [60] model using the Miche criterion with a breaking parameter of 0.8 [61].

SISYPHE is used to parametrise transport estimates from input from the hydrodynamic and wave models and to then balance the transport and bed evolution with the Exner equation. For simplicity and efficiency, the Soulsby Van Rijn [62] formulations were chosen for this study because, among other things, they can be used to simply estimate the suspended-load contribution to the bed-load components of the total sand transport. To replicate the cross-shore varying sediment grain size described in [55], three sediment classes are defined. The populations of the three sediment classes were divided into two zones at the 12 m-depth contour. In waters shallower than 12 m, there were 50% fine sediment diameters ($d_{50,1}$ = 0.3 mm) and 50% medium ($d_{50,2}$ = 0.4 mm) sand grain sizes. In waters deeper than 12 m, there were 50% medium ($d_{50,2}$ = 0.4 mm) and 50% coarse ($d_{50,3}$ = 0.8 mm) sand grain sizes. Summing the sediment populations resulted in an initial D_{50} of 0.035 mm in waters less than 12 m and 0.6 mm in waters deeper than 12 m. A second combination of sediment classes was tested, but showed small internal model sensitivity compared to the different wave breaking parameterisation [51].

In the calibration process, the model was found to underestimate the surveyed seabed evolution. This difference could be explained by the use of constant horizontal viscosity $\subseteq_t$ = 10^{-6} across the model domain, where field studies have measured a value in the vertical direction ~100 times greater within the surf zone [63]. Kinematic viscosity $\subseteq$ has also been noted to be 100 times larger when there is sediment in the water column during large wave events [16]. With this underestimation of $\subseteq$, the dimensionless grain size (D_*) in the Soulsby Van Rijn [62] formulae would be overestimated, leading to an underestimation of the suspended transport by a multiple of ~30 during large wave events. The increased eddy viscosity at the bed level would also impact bed load transport, which cannot be accounted for in the Soulsby Van Rijn transport equations. Hence, total sediment transport is scaled up to improve the model's ability to replicate the bed evolution surveys and represent an increased viscosity/diffusivity in the sediment transport model.

4.2. Change Factor Downscaling

GCM derived monthly change values were identified in [5]. Normalised climate monthly mean (accented with a bar over the variable) model anomalies ($\overline{A}$) were calculated as the difference between the monthly mean future ($\overline{F_G}$) 2081–2100 and baseline ($\overline{B_G}$) 1981–2000 GCM model output relative to the standard deviation of the baseline ($sd(\overline{B_G})$) output, where

$$\overline{A} = \left(\overline{F_G} - \overline{B_G}\right)/sd\left(\overline{B_G}\right). \tag{1}$$

The downscale CF for the future longshore current forcing was found by rearranging Equation (1) and replacing the monthly climate model baseline mean ($\overline{B_G}$) with the hourly hindcast longshore current data (B_{HU}) and calculating the hindcast monthly mean standard deviation ($sd(\overline{B_{HU}})$). The boundary conditions for future longshore currents (Figure 3) are therefore defined by the following equation;

$$F_U = \overline{A}\cdot sd\left(\overline{B_{HU}}\right) + B_{HU}, \tag{2}$$

where F_U is the hourly time series of the imposed future longshore current speed and $\overline{A}$ is the GCM derived mean normalised anomaly for a given month [5]. The future current speeds were scaled to the hindcast longshore current vector on the prescribed flow boundary.

For waves, the CERC longshore wave transport equation was used for the CF downscaling (Equation (1)) and was sourced again from [5]. The CERC equation is a function of both the breaking wave height (H_b) and breaking incident wave direction (θ_{ib}) [15]. The CERC equation is written as;

$$Q_w(H_b, \theta_{ib}) = K_s H_b^{5/2} \sin(2\theta_{ib})/2,$$
$$K_s = \frac{0.023 g^{1/2}}{(s-1)},$$

(3)

where $g = 9.8$ m/s^2 is acceleration due to gravity and $s = 2.6$ is the ratio of sediment and water densities. Wave breaking parameters were not available, so H_b was assumed to have the value of the nearshore H_s. θ_{ib} was calculated from the nearshore mean wave direction (θ) clockwise from true north so that θ_i equals $\theta - \theta_N$, where θ_N is the angle of the shore normal.

2D wave spectra changes (with frequency and direction dimensions) were not available for GCM derived TELEMAC simulations. The GCM normalised wave change analysis was summarised by the single bulk CERC parameter of Q_w, which is a function of two parameters, H_s and θ_i [5]. For simplicity in the TOMAWAC model setup it is assumed that the change in the GCM derived CERC Q_w can be entirely defined by an offset in wave climate direction alone. We define the wave-transport-directional CF (θ_Δ) term to describe more than the simple change in the mean wave direction, as it takes into account changes in wave height to modify the longshore transport. For the future longshore wave transport, Equation (3) is written as the following;

$$F_Q = Q_w\left(H_{sH}, \theta_{iH} + \overline{\theta_\Delta}\right),$$

(4)

where F_Q is the hourly time series of future mean longshore wave transport resulting from the hourly input from the wave hindcast (H_{sH}, θ_{iH}) and the monthly mean offset in wave-transport-directional CF ($\overline{\theta_\Delta}$). Future longshore wave transport can be solved a second way. By using Equation (1) again, we can solve for the CF change applied to the hindcast;

$$F_Q = \overline{A} \cdot \text{sd}\left(Q_w(H_{sH}, \theta_{iH})\right) + Q_w(H_{sH}, \theta_{iH})$$

(5)

where A is the GCM-derived mean normalised anomaly for a given month from Equation (1), $\text{sd}\left(Q_w(H_{sH}, \theta_{iH})\right)$ is the standard deviation of the monthly-mean longshore transport across each of the ~30 years. The monthly values of θ_Δ (Equation (4) substituted into Equation (3)) were found numerically by minimising (optimising) the absolute error between Equations (4) and (5). The optimisation method was implemented using the statistical software package R [64].

To generate the future rotated 2D wave spectrum boundary condition (Figure 3), the monthly mean $\overline{\theta_\Delta}$ was subtracted from the direction coordinate of the hindcast 2D wave spectrum. This spectrum (with the shifted direction coordinates) was then interpolated back onto the original direction coordinates.

Future climate simulations forced with sea level rise were achieved by simply adding the value of sea level increase to the boundary free surface water level data (Figure 3). Estimates of global mean sea level rise for the period 2081–2100—compared to 1986–2005—is likely (medium confidence) to be in the 5% to 95% range of projections from process-based models, which give 0.26 to 0.55 m for RCP2.6 and 0.45 to 0.82 m for RCP8.5 [65]. For RCP8.5, the rise by 2100 is 0.52 to 0.98 m with a rate during 2081–2100 of 8 to 16 mm year^{-1}. Increasing the water level by an additional one metre would result in a separation of the bottom sediment transport from the water surface waves within the 3.7 year climate simulation, whereas the natural climate sea level rise would gradually change over decades. Instead, future simulations included an instantaneous increase in sea level rise, which is less than the tidal amplitude.

4.3. Semi-Empirical NMB-LM Equation

The empirical CERC formulation (Equation (3)) was reformulated to include the effect of longshore wind- and tide-driven currents to better represent the transport modelled by TELEMAC. Once calibrated, the updated CERC equation (NMB-LM) was used to extrapolate the ~3.7 year TELEMAC hindcast simulation to the full ~30 year WW3 and ROMS hindcast datasets, to get a larger picture of the transport climate estimated by TELEMAC. It was then used with CF input wave and flow parameters to estimate the effect of climate change on the longer ~30 year hindcast. The reformulated empirical flow equation was composed of the following;

$$Q_{wu} = D_1 K_s H_s^{D_2} \frac{\sin(2\theta_i D_3)}{2} + D_4 U_l |U_l| \left| D_1 K_s H_s^{D_2} \frac{\sin(2\theta_i D_3)}{2} \right|, \tag{6}$$

where D_1, D_2, D_3 and D_4 are the calibrated values and U_l is the longshore current velocity, positive (negative) eastward (westward). The parameters of D were found by minimising (optimising) the absolute error between the simulated longshore transport of TELEMAC (with 2D spectral input) and Equation (6) (with the bulk WW3 parameters). The optimisation method of [64] implemented in the statistical software package R, was used.

The first term on the right of Equation (6) is the CERC equation scaled by D_1 and has H_s changed from the power of 5/2 to D_2. Different expressions of the power of Hs have been explained by [13,66]. The power of H_s was increased because the CERC equation predicted more transport than TELEMAC during low wave conditions compared to large wave conditions. The incident angle (θ_i) is also scaled down by D_3 because the offshore waves input in ~20 m of water has a greater incident angle to the coast than the required CERC wave breaking angle. The second term on the right of Equation (6) is the product of the longshore current U_l, a scale parameter D_3, the absolute value (magnitude) of U_l and the absolute value of the first term on the right (scaled CERC wave term). This formulation replicated the TELEMAC-derived transport, where the longshore flow (U_l) only had a significant effect when the wave transport is large.

5. Results

In the following three subsections, validation is provided for the TELEMAC sediment transport model, followed by comparison of the transport estimates from the TELEMAC model and the CERC and novel NMB-LM empirical equations and then the results from the novel downscaled method for climate change projections are presented respectively. Supplementary information is described in detail in the supporting dissertation paper [51].

5.1. TELEMAC Sediment Transport Model Validation

Morphodynamic model validation was undertaken by comparing the measured elevation change to the modelled bathymetric evolution between single-beam surveys for the first nine simulations (10 surveys over 33 months). The bathymetric change was compared between the model and the surveys in the vicinity of the storm trough (100–200 m from the coast) and storm bar (200–300 m from the coast) at both the west and east DMGs (Figure 1). The storm bar and trough are the most active features in the bathymetry. The calibrated model does a reasonable job of capturing the magnitude and direction (down/erosion or up/accretion) of bed evolution of the storm bar and trough in all but a few surveys (Figure 4). Any differences between model and observations could be a result of dredge disposal between surveys, the effect of the channel, or from under resolving the sediment transport forcing (e.g., 3D effects). The validation statistics on all data points results in a Pearson's R-value of 0.9, a P-value 6.2×10^{-14} and a standard error of 0.068 m.

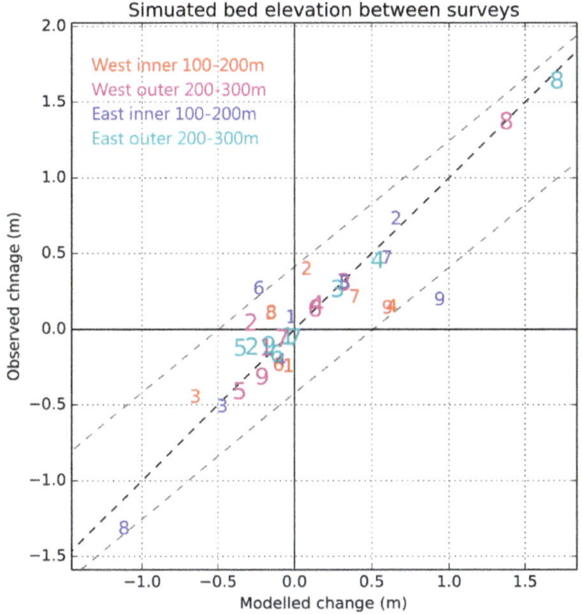

Figure 4. Model bed evolution validation in the location of the storm-bar and trough at both single-beam survey sites. Plotted are the model bed elevation median changes over the first nine simulations (10 surveys over 33 months). The dashed grey line is plus or minus one standard deviation of a linear fit to all data. The dashed black line is the one-to-one line.

5.2. Model Comparison

The time series plot of modelled longshore transport (Figure 5) shows that in the periods between storms, the CERC equation predicts more gross transport than the TELEMAC model. The same plot shows that during storms, the CERC equation predicts less transport than TELEMAC. Over the ~3.7 year period the cumulative/net effect of larger prediction by the CERC equation between storms and lower prediction during storms, balances out to match the TELEMAC storm-driven transport (Table 1). The reason for the difference between CERC and TELEMAC is the modelling regime, i.e., TELEMAC limits transport to occur only when the non-linear combination of coastal flow and orbital wave velocity is above a critical mobility velocity in the Soulsby Van Rijn equations. On the other hand, the CERC equation will predict transport for all wave heights and non-zero incident wave directions. The CERC equation is based on the work of [12] and was designed for a wave-dominated coast. The newly developed semi-empirical equation (NMB-LM) used to include the effect of currents modelled by TELEMAC is detailed in Section 4.3. The calibrated values of the NMB-LM model (Equation (6)) to the TELEMAC simulations are $D_{i=1:4}$ = {0.30625, 5.65716, 0.07662, 2.77079}. NMB-LM matches the pattern of net longshore transport modelled by TELEMAC, differing in magnitude for only a few storm events (Figure 5).

The ~30 year hindcast forcing was applied to the NMB-LM and CERC models to predict the long-term annual transport rate (Table 1). The CERC equation predicts a similar annual transport modelled to the ~3.7 year bathymetric survey period, however the NMB-LM equation estimates only 50% of the transport in the shorter period because of the storm-driven transport modelling regime.

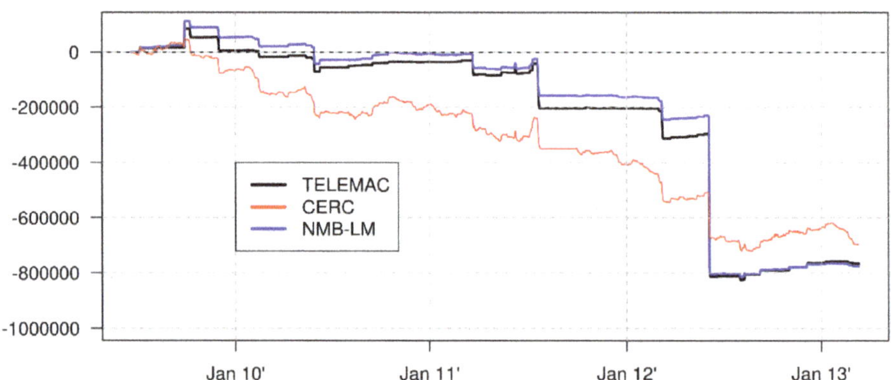

Figure 5. Time series of net longshore transport (m^3) from TELEMAC and empirical equations. On the vertical axis, positive (negative) values represent transport in the eastward (westward) direction.

Table 1. Net modelled longshore sediment transport rate Q estimates averaged per year.

Transport Per Year (m^3 m^{-1} year^{-1})	TELEMAC	CERC	NMB-LM
~3.7 year TELEMAC survey simulation period	−211,598	−193,379	−214,308
~30 year hindcast ROMs period	Not resolved.	−206,494	−100,741

5.3. Downscaled Climate Change Analysis

Prior to analysing the wave transport change, the wave-transport-directional CF was first checked for correctness. The CERC-derived wave transport climate modelled with the wave-transport-directional CF in Equation (4), resulted in the same wave transport climate as that modelled with the normalised GCM wave transport anomaly and Equation (5). This confirms that the wave-transport-directional CF can account for the GCM-derived changes in wave transport. In other words, the wave-transport-directional CF can account for the added dependence in the changes in wave height. Also, the wave CF factor applied to the NMB-LM equation (Equation (6)) and then used to rotate the TELEMAC directional spectrum resulted in similar wave change climates.

The time series of baseline, net longshore sediment transport (m^3) climate (Figure 5), is replotted with the five CF climate sensitivity runs in Figure 6. It is difficult to numerically model a steady sediment transport (and bed evolution) simulation over long time scales, because of cumulative errors. Therefore, it is difficult to simulate the period of predicted gradual increased sea level rise by the end of the century (i.e., a 100 year simulation) to around 0.8–1.0 m predicted by high GHG future simulations. To stabilise the TELEMAC climate sensitivity simulations, the bathymetry was reset to the measured profiles to be consistent with the baseline simulations. The modelling here showed little impact (1.3%–2.1% change) on the longshore transport, for relatively large changes in sea level (0.1–0.2 m over ~3.7 years). The nonlinear influence of the increase in sea level means it difficult to extrapolate these changes out to the end of the 21st century (0.8–1.0 m) levels.

The TELEMAC and NMB-LM modelled transport rates are dependent on wind-tide-driven currents. The CF climate analysis in Figure 6 showed small sensitivity (1–2%) from the impact of changes from the wind-driven current forcing.

The TELEMAC derived wave CF simulation are shown to be the primary driver of change to future longshore transport (Figure 6). The wave CF projects a 51.1% decrease in net westward longshore transport. Over the same period the CERC and NMB-LM predicted a similar 49% and 46.6% decrease, respectively.

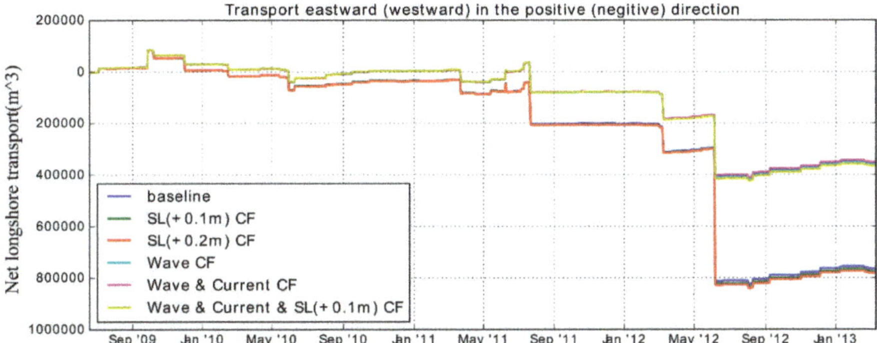

Figure 6. Time series of net longshore transport $Q_T\delta$ (m^3) from different TELEMAC downscaled climate change factor (CF) forcing. This plot has the same format as Figure 5. The legend indicates which change factor (CF) sensitivity forcing has been applied, along with the baseline transport. SL is the sea level increase (0.1 or 0.2 m).

The CERC and NMB-LM equations were used to estimate the four member ensemble model spread from the different GCM-derived wave transport changes [5]. The HadGEM2-ES predicts the largest change (95.7%), and the INMCM4 (−14.5%) the smallest change, but all projections suggest a decrease in the baseline westward transport by around 50% (±40%) [51].

Further investigation into the monthly contribution to the annual change value indicates June is the dominant contributor to the annual change [51]. Changes over the shorter modelled TELEMAC simulation period are similar to the CERC and NMB-LM climates for the same period and the longer hindcast period. However, analysis of the monthly contribution to the change suggests that shorter sediment transport simulations (3–5 years) are unable to accurately represent the longer climate (20 years) monthly change signal.

6. Conclusions and Future Work

This study presents a method for downscaling the GCM derived drivers of transport changes to investigate the sediment transport at a location that may be influenced by projected global circulation changes. Much of the analysis of future wave-climate change is somewhat limited to studies focusing on the parameter of significant wave height [43]. A new method is introduced to downscale GCM longshore sediment transport projections using the wave-transport-directional change parameter to modify directional wave spectra. The directional wave CF method was applied to the CAWCR hindcast in the TELEMAC simulations, checked against the empirically downscaled climate, and provided the same projected monthly transport climate results. The advantage the CF method has over the dynamically downscaled method is that the baseline variability and representation of extremes are better represented in the hindcast CF dataset than in the coarse resolution GCM datasets.

The TELEMAC coastal area-type model, CERC and new NMB-LM coastline-type models, predict a similar annual net longshore sediment transport of ~200,000 m^3 year^{-1} westward over the bathymetric ~3.7 year survey period. Over the longer ~30 year wave and hydrodynamic hindcast period CERC predicts similar annual transport to the ~3.7yr period, however, the NMB-LM predicts around half of the transport, ~100,000 m^3 year^{-1}. This is a similar value to the one used in the dredging program for the local port authority [24].

The CF downscaled sediment transport modelling predicts around a 50% (±40%) decrease in the westward net longshore sediment transport. The TELEMAC sediment transport climate simulations, predict a non-linear effect to increases in water levels (0.1 and 0.2 m). The sensitivity of the climate projections from the influence of increased sea level, results in only a 1% to 2% change in the transport.

Projected changes to wind-driven currents also resulted in a 1% to 2% change in the future projection of transport.

The main driver of the identified change is from the influence of wave-driven transport changes. The contribution from changes in individual months was also shown to be important [51]. While large normalised current change was captured in [5]—particularly in summer months when the subtropical ridge location (STR-L) is over NMB—the change associated with waves during the largest winter storms drive the overall transport climate. It is therefore difficult to directly associate the remote winter time STR-L position located to the north of NMB, with the overall identified transport change.

The TELEMAC modelled transport rates presented are dependent on wind-tide-driven currents. However the climate analysis showed small sensitivity (1% to 2%) from the timing, or normalised change, from the wind-driven currents. This study provides evidence of the weak secondary importance of changes to wind-driven currents compared to the dominant wave-driven transport in analysing the average change in sediment transport climate at NMB.

Sediment transport models have been applied to the nearshore coastal region since the 1980s. Today, there are several advanced numerical-based coastal area-type models, which are available to model sediment transport. Possible future additions to these models, are Lagrangian flow coordinates or adaptive grid-mesh. Arguably, more development is required on sub-grid scale parametrisations of processes such as wave asymmetry/skewness [67,68], bottom roughness [69–72]) and viscosity and diffusivity in the surf zone [16,63]. This requires more empirical measurements, which further highlights the need for continued measurements for the dredging program at Lakes Entrance.

Bringing together the modelling improvements, the ultimate aim is to be able to model complex, long-term, decadal sediment transport simulations and to run models that are able to resolve the variability and change, in the advance or retreat of the coastline position.

Author Contributions: Conceptualization, J.O.; formal analysis, J.O.; investigation, J.O.; methodology, J.O.; supervision, A.B. and K.M.; validation, J.O.; writing—original draft, J.O.; writing—review and editing, A.B. and K.M.

Funding: This study was supported by the Earth Science and Climate Change Hub of the Australian Government's National Environmental Science Programme (NESP). This work also forms part of the lead author's PhD thesis.

Acknowledgments: We would like to thank the two anonymous journal reviewers for their valuable comments that have improved this manuscript. The lead author would like to thank Mark Spykers from Gippsland Ports for the use of the observational datasets and Mark Hemer, Claire Trenham and Frank Colberg for access to the modelled datasets.

Conflicts of Interest: The authors declare no conflict of interest.

References

1. Wong, P.; Lonsada, I.; Gattuso, J.; Hinkel, J.; Burkett, V.; Codignotto, J. Coastal Systems and Low-Lying Areas. In *Climate Change 2014 Impacts, Adaptation and Vulnerability*; Cambridge University Press: Cambridge, UK, 2014; pp. 361–410. ISBN 00472425.
2. Bruun, P. Sea-level rise as a cause of shore erosion. *J. Geol.* **1962**, 76–92.
3. Aagaard, T.; Sørensen, P. Sea level rise and the sediment budget of an eroding barrier on the Danish North Sea coast. In Proceedings of the 12th International Coastal Symposium, Plymouth, UK, 8–12 April 2013; pp. 434–439.
4. Houston, J.R.; Dean, R.G. Shoreline Change on the East Coast of Florida. *J. Coast. Res.* **2014**, *30*, 647–660.
5. O'Grady, J.G.; McInnes, K.L.; Colberg, F.; Hemer, M.A.; Babanin, A.V.; O'Grady, J.G.; McInnes, K.L.; Colberg, F.; Hemer, M.A. Longshore Wind, Waves and Currents: Climate and Climate Projections at Ninety Mile Beach, South-eastern Australia. *Int. J. Climatol.* **2015**, *35*, 4079–4093. [CrossRef]
6. Russell, K.; Rennie, J.; Sjerp, E. *Gippsland State of the Coast Update*; Water Technology: Melbourne, Australia, 2013.
7. Hemer, M.A.; McInnes, K.L.; Ranasinghe, R. Climate and variability bias adjustment of climate model-derived winds for a southeast Australian dynamical wave model. *Ocean Dyn.* **2011**, *62*, 87–104. [CrossRef]

8. Erikson, L.H.; Hegermiller, C.A.; Barnard, P.L.; Ruggiero, P.; van Ormondt, M. Projected wave conditions in the Eastern North Pacific under the influence of two CMIP5 climate scenarios. *Ocean Model.* **2015**, *96*, 171–185. [CrossRef]

9. Mil-homens, J.; Ranasinghe, R.; De Vries, J.V.T.; Stive, M.J.F. Re-evaluation and improvement of three commonly used bulk longshore sediment transport formulas. *Coast. Eng.* **2013**, *75*, 29–39. [CrossRef]

10. Tomasicchio, G.R.; Alessandro, F.D.; Barbaro, G.; Malara, G. General longshore transport model. *Coast. Eng.* **2013**, *71*, 28–36. [CrossRef]

11. Bayram, A.; Larson, M.; Hanson, H. A new formula for the total longshore sediment transport rate. *Coast. Eng.* **2007**, *54*, 700–710. [CrossRef]

12. Komar, P.D. The mechanics of sand transport on beaches. *J. Geophys. Res.* **1971**, *76*, 713. [CrossRef]

13. Kamphuis, J.W.; Davies, M.H.; Nairn, R.B.; Sayao, O.J. Calculation of Littoral Sand Transport Rate. *Coast. Eng.* **1986**, *10*, 1–21. [CrossRef]

14. van Rijn, L.C. A simple general expression for longshore transport of sand, gravel and shingle. *Coast. Eng.* **2014**, *90*, 23–39. [CrossRef]

15. CERC. U.S. Army corps of Engineers shore protection manual-Volume I. *Coast. Eng. Res. Cent.* **1984**, *1*, 652.

16. Van Rijn, L.C. Simple general formulae for sand transport in rivers, estuaries and coastal waters. Available online: www.leovanrijn-sediment.com (accessed on 11 November 2017).

17. Roelvink, D.; Reniers, A. *A Guide to Modelling Coastal Morphology*; World Scientific: Singapore, 2011.

18. Longuet-Higgins, M.S.; Stewart, R.W. Radiation stresses in water waves; a physical discussion, with applications. *Deep Sea Res.* **1964**, *11*, 529–562. [CrossRef]

19. Longuet-Higgins, M.S. Longshore currents generated by obliquely incident sea waves: 1. *J. Geophys. Res.* **1970**, *75*, 6778. [CrossRef]

20. van Rijn, L.C.; Walstra, D.-J.R.; van Ormondt, M. Unified View of Sediment Transport by Currents and Waves. IV: Application of Morphodynamic Model. *J. Hydraul. Eng.* **2007**, *133*, 776–793. [CrossRef]

21. Vitousek, S.; Barnard, P.L.; Limber, P.; Erikson, L.; Cole, B. A model integrating longshore and cross-shore processes for predicting long-term shoreline response to climate change. *J. Geophys. Res. Earth Surf.* **2017**, *122*, 782–806. [CrossRef]

22. Ashton, A.; Murray, A.B.; Arnault, O. Formation of coastline features by large-scale instabilities induced by high-angle waves. *Nature* **2001**, *414*, 296–300. [CrossRef] [PubMed]

23. Szmytkiewicz, M.; Biegowski, J.; Kaczmarek, L.M.; Okroj, T.; Ostrowski, R.; Pruszak, Z.; Różyńsky, G.; Skaja, M. Coastline changes nearby harbour structures: Comparative analysis of one-line models versus field data. *Coast. Eng.* **2000**, *40*, 119–139. [CrossRef]

24. GHD. *Gippsland Lakes Ocean. Access Program. 2012 TSHD Maintenance Program. Final*; Gippsland Ports: Bairnsdale, Australia, 2013.

25. Samaras, A.G.; Gaeta, M.G.; Miquel, A.M.; Archetti, R. High-resolution wave and hydrodynamics modelling in coastal areas: Operational applications for coastal planning, decision support and assessment. *Nat. Hazards Earth Syst. Sci.* **2016**, *16*, 1499–1518. [CrossRef]

26. Putzar, B.; Malcherek, A.M. Development of a long-term morphodynamic model for the German Bight. In Proceedings of the 19th Telemac-Mascaret User Conference, Oxford, UK, 18–19 October 2012; pp. 47–52.

27. Chini, N.; Stansby, P. Broad-Scale Hydrodynamic Simulation, Wave Transformation and Sediment Pathways. In *Broad Scale Coastal Simulation: New Techniques to Understand and Manage Shorelines in the Third Millennium*; Nicholls, R.J., Dawson, R.J., Day, S.A., Eds.; Springer: Dordrecht, The Netherlands, 2015; pp. 103–124. ISBN 978-94-007-5258-0.

28. Fowler, H.J.; Blenkinsop, S.; Tebaldi, C. Linking climate change modelling to impacts studies: Recent advances in downscaling techniques for hydrological modelling. *Int. J. Climatol.* **2007**, *27*, 1547–1578. [CrossRef]

29. Ekström, M.; Grose, M.R.; Whetton, P.H. An appraisal of downscaling methods used in climate change research. *Wiley Interdiscip. Rev. Clim. Chang.* **2015**, *6*, 301–319. [CrossRef]

30. Weisse, R.; von Storch, H. Past and future changes in wind, wave, and storm surge climates. In *Marine Climate and Climate Change*; Springer: Berlin/Heidelberg, Germany, 2010; pp. 165–203.

31. Katzfey, J.J.; McGregor, J.; Nguyen, K.; Thatcher, M. Dynamical downscaling techniques: Impacts on regional climate change signals. In Proceedings of the 18th World IMACS Congress and MODSIM09 International Congress on Modelling and Simulation, Cairns, Australia, 13–17 July 2009; pp. 3942–3947.

32. Kendon, E.J.; Ban, N.; Roberts, N.M.; Fowler, H.J.; Roberts, M.J.; Chan, S.C.; Evans, J.P.; Fosser, G.; Wilkinson, J.M. Do convection-permitting regional climate models improve projections of future precipitation change? *Bull. Am. Meteorol. Soc.* **2017**, *98*, 79–93. [CrossRef]

33. Hemer, M.A.; McInnes, K.L.; Ranasinghe, R. Projections of climate change-driven variations in the offshore wave climate off south eastern Australia. *Int. J. Climatol.* **2013**, *33*, 1615–1632. [CrossRef]

34. Vousdoukas, M.I.; Voukouvalas, E.; Annunziato, A.; Giardino, A.; Feyen, L. Projections of extreme storm surge levels along Europe. *Clim. Dyn.* **2016**, *47*, 3171–3190. [CrossRef]

35. Erikson, L.H.; Hemer, M.A.; Lionello, P.; Mendez, F.J.; Mori, N.; Semedo, A.; Wang, X.L.; Wolf, J. Projection of wave conditions in response to climate change: A community approach to global and regional wave downscaling. In *The Coastal Sediments 2015*; World Scientific: Singapore, 2015; p. 14. ISBN 9789814689977.

36. Wandres, M.; Pattiaratchi, C.; Hemer, M.A. Projected changes of the southwest Australian wave climate under two atmospheric greenhouse gas concentration pathways. *Ocean. Model.* **2017**, *117*, 70–87. [CrossRef]

37. Charles, E.; Idier, D.; Delecluse, P.; Déqué, M.; Le Cozannet, G.; Cozannet, G. Climate change impact on waves in the Bay of Biscay, France. *Ocean. Dyn.* **2012**, *62*, 831–848. [CrossRef]

38. Howard, T.; Lowe, J.; Horsburgh, K. Interpreting century-scale changes in southern north sea storm surge climate derived from coupled model simulations. *J. Clim.* **2010**, *23*, 6234–6247. [CrossRef]

39. Duong, T.M.; Ranasinghe, R.; Thatcher, M.; Mahanama, S.; Wang, Z.B.; Dissanayake, P.K.; Hemer, M.; Luijendijk, A.; Bamunawala, J.; Roelvink, D.; et al. Assessing climate change impacts on the stability of small tidal inlets: Part 1—Data poor environments. *Mar. Geol.* **2017**, *390*, 331–346. [CrossRef]

40. Duong, T.M.; Ranasinghe, R.; Thatcher, M.; Mahanama, S.; Wang, Z.B.; Dissanayake, P.K.; Hemer, M.; Luijendijk, A.; Bamunawala, J.; Roelvink, D.; et al. Assessing climate change impacts on the stability of small tidal inlets: Part 2—Data rich environments. *Mar. Geol.* **2018**, *395*, 65–81. [CrossRef]

41. Mortlock, T.R.; Goodwin, I.D.; McAneney, J.K.; Roche, K. The June 2016 Australian East Coast Low: Importance of wave direction for coastal erosion assessment. *Water* **2017**, *9*, 121. [CrossRef]

42. Casas-Prat, M.; McInnes, K.L.; Hemer, M.A.; Sierra, J.P. Future wave-driven coastal sediment transport along the Catalan coast (NW Mediterranean). *Reg. Environ. Chang.* **2016**, *16*, 1739–1750. [CrossRef]

43. Morim, J.; Hemer, M.; Cartwright, N.; Strauss, D.; Andutta, F. On the concordance of 21st century wind-wave climate projections. *Glob. Planet. Change* **2018**, *167*, 160–171. [CrossRef]

44. Zhang, X.; Church, J.A.; Monselesan, D.; McInnes, K.L. Sea level projections for the Australian region in the 21st century. *Geophys. Res. Lett.* **2017**, *44*, 8481–8491. [CrossRef]

45. Caires, S.; Swail, V.R.; Wang, X.L. Projection and analysis of extreme wave climate. *J. Clim.* **2006**, *19*, 5581–5605. [CrossRef]

46. Von Storch, H.; Reichardt, H. A scenario of storm surge statistics for the german bight at the expected time of doubled atmospheric carbon dioxide concentration. *J. Clim.* **1997**, *10*, 2653–2662. [CrossRef]

47. Antolínez, J.A.A.; Murray, A.B.; Méndez, F.J.; Moore, L.J.; Farley, G.; Wood, J. Downscaling Changing Coastlines in a Changing Climate: The Hybrid Approach. *J. Geophys. Res. Earth Surf.* **2018**, *123*, 229–251. [CrossRef]

48. Anandhi, A.; Frei, A.; Pierson, D.C.; Schneiderman, E.M.; Zion, M.S.; Lounsbury, D.; Matonse, A.H. Examination of change factor methodologies for climate change impact assessment. *Water Resour. Res.* **2011**, *47*, 1–10. [CrossRef]

49. Katzfey, J.J.; McInnes, K.L. GCM simulations of eastern Australian cutoff lows. *J. Clim.* **1996**, *9*, 2337–2355. [CrossRef]

50. McInnes, K.L.; Hubbert, G.D. The impact of eastern Australian cut-off lows on coastal sea levels. *Meteorol. Appl.* **2001**, *8*, 229–243. [CrossRef]

51. O'Grady, J.G. Nearshore Modelling of Longshore Sediment Transport in the Application to Climate Change Studies at Ninety Mile Beach, Australia. Ph.D. Thesis, Swinburne University of Technology, Melbourne, Australia, 2018.

52. Durrant, T.; Greenslade, D.; Hemer, M.A.; Trenham, C. *Global Wave Hindcast Focussed on the Central and South. Pacific*; CAWCR: Aspendale, Australia, 2014.

53. Colberg, F.; McInnes, K.L.; O'Grady, J.; Hoeke, R.K. Atmospheric Circulation Changes and their Impact on Extreme Sea Levels around Australia. *Nat. Hazards Earth Syst. Sci.* **2019**, *19*, 1067–1086. [CrossRef]

54. Sinclair, M.J.; Quadros, N. *Airborne Lidar Bathymetric Survey for Climate Change Airborne Lidar Bathymetric Survey for Climate Change*; Springer: Berlin/Heidelberg, Germany, 2010; pp. 11–16.

55. Wright, L.D.; Nielsen, P.; Short, A.D.; Coffey, F.C.; Green, M.O. *Nearshore and Surfzone Morphodynamics of a Storm Wave Environment Eastern Bass Strait*; Sydney Univ, Australian Coastal Studies Unit: Sydney, Australia, 1982.

56. Hervouet, J.-M. *Hydrodynamics of Free Surface Flows: Modelling With the Finite Element Method*; John Wiley & Sons: Hoboken, NJ, USA, 2007.

57. Villaret, C.; Hervouet, J.-M.M.; Kopmann, R.; Merkel, U.; Davies, A.G. Morphodynamic modeling using the Telemac finite-element system. *Comput. Geosci.* **2013**, *53*, 105–113. [CrossRef]

58. Saint-Venant, B.D. Theory of unsteady water flow, with application to river floods and to propagation of tides in river channels. *French Acad. Sci.* **1871**, *73*, 237–240.

59. Benoit, M.; Marcos, F.; Becq, F. Development of a third generation shallow-water wave model with unstructured spatial meshing. *Coast. Eng. Proc.* **1996**, 465–478.

60. Battjes, J.A.; Janssen, J.P.F.M. Energy Loss and Set-Up Due To Breaking of Random Waves. *Coast. Eng. Proc.* **1978**, 569–587.

61. Miche, A. Mouvements ondulatoires de la mer en profondeur croissante ou decroissante. *Ann. des Ponts Chaussees* **1944**, *144*, 369–406.

62. Soulsby, R.L. *Dynamics of Marine Sands*; Thomas Telford: London, UK, 1997.

63. Wright, L.D.; Nielsen, P.; Shi, N.C.; List, J.H. Morphodynamics of a bar-trough surf zone. *Mar. Geol.* **1986**, *70*, 251–285. [CrossRef]

64. Nelder, J.A.; Mead, R. A simplex algorithm for function minimization. *Comput. J.* **1965**, *7*, 308–313. [CrossRef]

65. Church, J.A.; Clark, P.U.; Cazenave, A.; Gregory, J.M.; Jevrejeva, S.; Levermann, A.; Merrifield, M.A.; Milne, G.A.; Nerem, R.; Nunn, P.D.; et al. Climate Change 2013: The Physical Science Basis, Contribution of Working Group I to the Fifth Assessment Report of the Intergovernmental Panel on Climate Change. In *Sea Level Change*; Cambridge University Press: Cambridge, UK, 2013; pp. 1137–1216.

66. Kamphuis, J.W. Alongshore Sediment Transport Rate. *J. Waterw. Port. Coastal Ocean. Eng.* **1991**, *117*, 624–640. [CrossRef]

67. Nielsen, P.; Callaghan, D.P. Shear stress and sediment transport calculations for sheet flow under waves. *Coast. Eng.* **2003**, *47*, 347–354. [CrossRef]

68. Van Rijn, L.; Ribberink, J.S.; van der Werf, J.J.; Walstra, D.J.R. Coastal sediment dynamics: Recent advances and future research needs. *J. Hydraul. Res.* **2013**, *51*, 475–493. [CrossRef]

69. Davies, A.G.; Villaret, C. Eulerian drift induced by progressive waves above rippled and very rough beds. *J. Geophys. Res.* **1999**, *104*, 1465–1488. [CrossRef]

70. Nielsen, P. Dynamics and geometry of wave-generated ripples. *J. Geophys. Res. Oceans* **2006**, *86*, 6467–6472. [CrossRef]

71. Smith, G.A.; Babanin, A.V.; Riedel, P.; Young, I.R.; Oliver, S.; Hubbert, G. Introduction of a new friction routine into the SWAN model that evaluates roughness due to bedform and sediment size changes. *Coast. Eng.* **2011**, *58*, 317–326. [CrossRef]

72. Villaret, C.; Huybrechts, N.; Davies, A.G.; Way, O. Effect of bed roughness prediction on morphodynamic modelling: Application to the Dee estuary (UK) and to the Gironde estuary (France). In Proceedings of the 34th IAHR World Congress, Brisbane, Australia, 26 June–1 July 2011; pp. 1149–1156.

Journal of
*Marine Science
and Engineering*

Article

Experimental Analysis of Wave Overtopping: A New Small Scale Laboratory Dataset for the Assessment of Uncertainty for Smooth Sloped and Vertical Coastal Structures

Hannah E Williams [1,*,†,‡], **Riccardo Briganti** [1,†], **Alessandro Romano** [2,†] **and Nicholas Dodd** [1,†]

1   Faculty of Engineering, University of Nottingham, Nottingham NG7 2RD, UK
2   Sapienza University of Rome, DICEA, via Eudossiana 18, 00184 Rome, Italy
*   Correspondence: h.e.williams@soton.ac.uk
†   These authors contributed equally to this work.
‡   Current Address: School of Geography and Environmental Science, University of Southampton, Southampton SO17 1BJ, UK.

Received: 14 June 2019; Accepted: 11 July 2019; Published: 13 July 2019

Abstract: Most physical model tests carried out to quantify wave overtopping are conducted using a wave energy spectrum, which is then used to generate a free surface wave time series at the wave paddle. This method means that an infinite number of time series can be generated, but, due to the expense of running physical models, often only a single time series is considered. The aim of this work is to investigate the variation in the main overtopping measures when multiple wave times series generated from the same spectrum are used. Physical model tests in a flume measuring 15 m (length) by 0.23 m (width) with an operating depth up to 0.22 m were carried out using a stochastic approach on two types of structures (a smooth slope and a vertical wall), and a variety of wave conditions. Results show variation of overtopping discharge, computed by normalising the range of the discharges at a certain wave condition with the maximum value of the discharge in the range up to 10%, when the same wave time series is used, but this range increases to 75% when different time series are used. This variation is found to be of a similar magnitude to both the one found with similar experiments looking at the phenomena in numerical models, and that specified by the confidence bounds in empirical methods.

Keywords: small scale physical modelling; wave overtopping; uncertainty analysis; random waves; wave spectra

---

1. Introduction

The coastal region has significant global economic and societal importance, leading to the extensive construction of coastal structures to defend these areas from the effects of wave action and currents. These structures are subjected to a variety of hydraulic processes, one of the most important and extensively studied being wave overtopping, due to the danger it poses to the people, property and infrastructure being protected. Therefore, the prediction of wave overtopping is crucial and it can be achieved using three established methods [1]: by using empirical formulae derived from large datasets of laboratory and field measurements (e.g., [2–5]); by using scaled physical model tests carried out in laboratories (e.g., [6–13]) some of which were included in the CLASH dataset [14] to calibrate formulae and neural network tools; or by simulating the hydraulic response of a structure using numerical models (e.g., [15–22]). Over the years, extensive research has been carried out into all of these methodologies; nonetheless, they are not perfect design tools, and they require further improvement.

Uncertainty in wave overtopping prediction has many sources. For example, even the methodology used by different laboratories introduces a variability in overtopping measures. This aspect was investigated by [14] and a complex method for dealing with it was introduced. Here, we are interested in one particular aspect that has recently gained attention, the quantification of the uncertainty induced by the generation of individual sequences of waves from the same energy density spectrum. This uncertainty occurs due to the fact that from every energy spectrum an infinite number of different waves sequences can be generated, simply by changing the initial seeding of the random phase distribution. Evidence that this plays an important role in the variability of results of numerical models was first given in studies by [23] for the prediction of run-up and [24] for overtopping which used a limited number of tests and waves, but showed that the parameters of overtopping can vary significantly with the seeding used. Subsequent work, using numerical models, by [25,26] carried out an extensive Monte Carlo analysis, and concluded that higher uncertainty was observed in both the overtopping discharge and the individual overtopping volumes when the probability of overtopping was less than 5%, with uncertainty decreasing as overtopping increased. Work by [27] also investigated a similar problem in the study of breakwater failure due to wave–structure interaction.

Uncertainty in the context of physical modelling was investigated in a number of works [6,8] with [6] first analysing this particular source of variability by performing a limited number of tests on a vertical wall, using different wave sequences generated from the same energy density spectrum. The work found that variability, due to the wave seeding, of the overtopping measured ranged from approximately between 8% and 15%, depending on the exact parameter of interest. Further work by [28] examined this same variability by considering a small set of laboratory experiments on a specific type of structure, namely a rubble mound breakwater. Similarly to numerical model results, it was shown that, when the probability of overtopping was less than 5% and for $R^* = R_c/H_{m0} > 2$ (where R_c is the structure crest freeboard above the still water level and H_{m0} is the spectral significant wave height at the structure toe), the variability was higher, decreasing as overtopping discharges increased. However, both of these studies were limited to a single structure type, which makes it impossible to fully extrapolate the results of the variability to other typical designs.

Although physical modelling is generally considered to be a reliable approach for the prediction of overtopping at coastal structures, when complex layouts or wave conditions are considered, the study by [28] identified that further research is required. A further motivation to the present study is given by the fact that this type of uncertainty in the laboratory tests is implicitly included in empirical formulae, which provide confidence bounds of the overtopping prediction to quantify uncertainty coming from all sources. Currently, it is not known how much of this uncertainty is generated simply by the seeding.

The purpose of this work is to address this gap by carrying out an extensive experimental analysis. A stochastic approach was used to quantify this source of uncertainty in physical models. Accordingly, a large number of overtopping experiments were carried out, by randomly altering the initial seeding values used to generate the wave time series for each experimental run. To increase the scope of this work, two different structure types were considered, one a smooth slope, and one a vertical wall, both representative of many existing coastal defence structures found around the world. In addition, a variety of wave conditions were considered, so that results can be extrapolated to different levels of overtopping.

As noted in [28], the study of uncertainty using multiple tests with different seeding forces to use small (i.e., with depth at generation of the order of a fraction of 1 m) scales. This inevitably introduces scale effects. Note that the experimental set-up in [28] is similar to the present one in terms of hydraulic parameters (e.g., water depth, wave characteristics.

When operating on a small scale, one of the most important effects on wave overtopping is given by the much lower air content in breaking waves interacting with the structure. A way of dealing with scale effects in overtopping prediction is provided by [5] in the form of a correction factor for the predicted overtopping discharge. However, it should be noted that, in the experiments presented in [28], when the variability of the results was analysed, scale effects acted in the same way in all tests; therefore, experiments can be compared among each other. However, the role of scale effects in affecting the variability remains unknown. Smooth structures were chosen so that the effect of the armour in terms of roughness and air entrainment was not present.

The paper is organised as follows; the following section describes the experimental set-up and instrumentation. Section 3 provides details about the test programme and experimental procedure, including information regarding the limiting of laboratory effects. Section 4 describes and analyses the effect that altering the seeding has on a number of overtopping parameters. Section 5 discusses the main points identified in the results; and finally Section 6 presents the overall conclusions of this research.

2. Experimental Set-Up

2.1. Flume Set-Up

The experiments were carried out in the wave flume at the University of Nottingham, which is approximately 15 m in length, 0.23 m in width and has an operating depth of up to 0.22 m. It is fitted with a piston type wave generator with active absorption (developed by HR Wallingford). The bottom of the flume is flat, but, for these experiments, a stainless steel foreshore was placed, starting at $x = 6.75$ m from the neutral position of the wave maker at $x = 0$ m. This allowed for the generation of depth-limited waves without the onset of wave breaking at the wave paddle. This foreshore was 4 m long in total, with a steeper 1 m section closest to the paddle end, followed by a 3 m section with a gradient of 1:50, which met the toe of the structures tested.

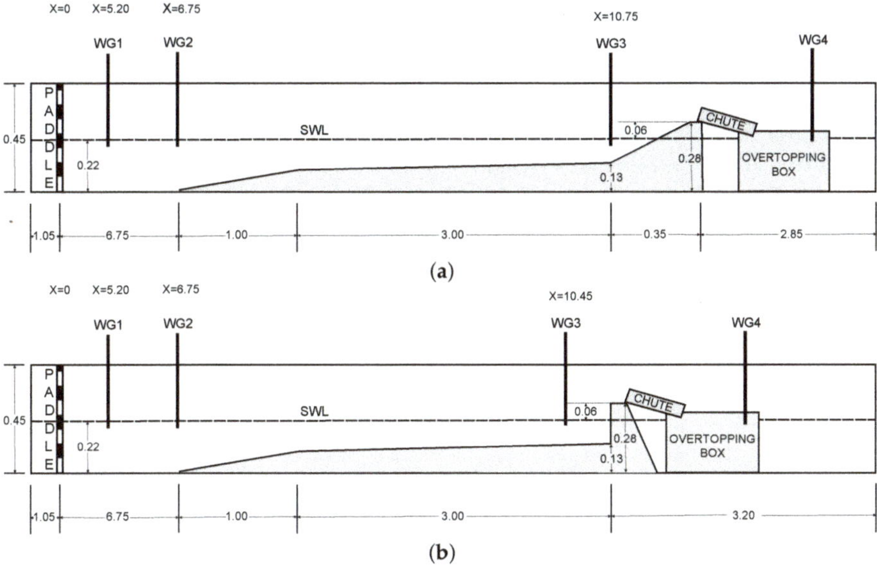

Figure 1. Indicative layouts of the physical models (not to scale). (**a**) smooth sloped structure; (**b**) vertical wall structure.

Two different overtoppable structures were independently placed in the flume on separate occasions for testing, namely an impermeable smooth slope with a gradient of 1:2.55 and a vertical wall. Both of the structures tested had an identical freeboard ($R_c = 0.06$ m). An overtopping tank was installed in the flume on the leeward side of the structures; water that overtopped the structures was directed into this tank. The full set-up of both experiments is detailed in Figure 1, with photographs of the two different structures shown in Figure 2.

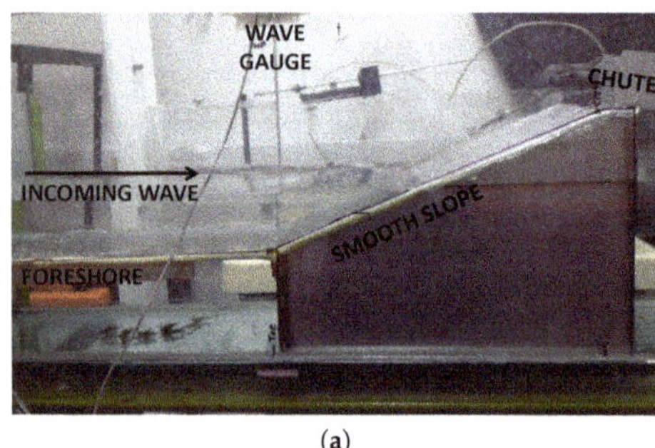

(a)

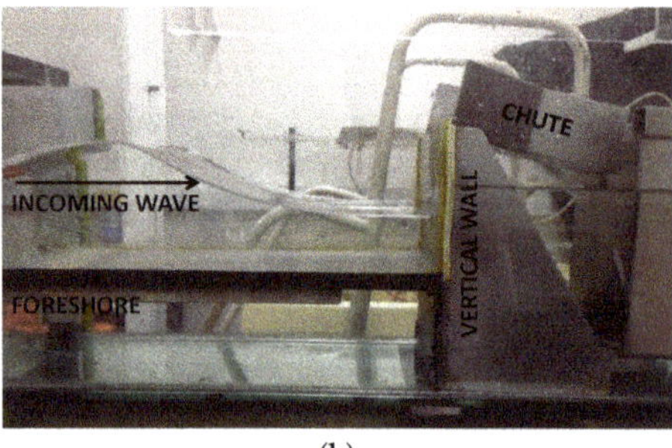

(b)

Figure 2. Photographs of the University of Nottingham experimental set-up. (**a**) smooth impermeable slope; (**b**) vertical wall.

2.2. Wave and Overtopping Measurement and Analysis

The water free surface (η) was measured along the flume using three wave gauges (*WG1–3*, see Figure 1 for their positions). *WG1* was located close to the paddle within the flat bed section, *WG2* was at the toe of the foreshore where waves were fully developed, and *WG3* was located close to the structure and was used to obtain the wave conditions at the toe. These wave gauges were of resistance type, and, due to their high sensitivity to laboratory conditions [29], they were calibrated twice daily during these experiments. This was to decrease the likelihood of the calibration affecting the reading, as water temperature changed throughout the day.

A standard procedure for measuring the overtopping volumes was used. A chute was placed just behind the crest of the structure. As water overtopped the structure, it entered this chute, and then flowed down into the overtopping tank situated at a distance behind the structure. The tank was constructed from stainless steel, measuring 0.5 m long by 0.22 m wide by 0.30 m deep, with a false wall in the centre, which allowed water to pass underneath, and dampened oscillations on the water surface in the rear section. Two chutes were used for these experiments, one of width 7.3 cm and one of width 22.0 cm, with the most appropriate one chosen depending on the expected discharge based on the empirical prediction. Using the wider chute for lower discharges allowed for more accurate measurements of the small individual events, whilst using the narrow chute for higher discharges removed the issue of the tank becoming full before the experimental run was complete. A wave gauge (*WG4*) was placed in the rear section of the tank to detect any changes in the water depth due to incoming wave overtopping. With the dimensions of both the chute and the box known for each experiment, these water depth changes were transformed into overtopping measures.

Changes in the water depth within the tank between the start and end of each experimental run were used to calculate the mean overtopping discharge, (q) for each experiment. To calculate the probability of overtopping (P_{ov}) and the individual overtopping volumes (V_{ov}), the time series of the individual overtopping volumes, i.e., the sudden changes in water depth multiplied by the surface area of the tank, was analysed. The original signal was filtered using a low pass digital filter. A peak detection method was then used to detect each individual overtopping event; the volumes were then calculated by working out the difference between subsequent peaks. However, due to the noise within this data, caused by oscillations in the water surface, there was still some residual uncertainty in the volume values calculated. Another limitation of this methodology that was followed was that it is quite possible that small overtopping events were not detected in the signal, hence lost, particularly in the high overtopping conditions, where multiple occurrences of overtopping took place in quick succession. This does not affect the discharge computations.

3. Test Programme

3.1. Incident Wave Conditions

This work is divided into two sets of experiments—one set carried out using the smooth slope structure, whilst the other set was carried out using the vertical wall. The irregular waves generated by the wave paddle in all tests used a JONSWAP spectra with a peak enhancement factor of $\gamma = 3.3$. For each structure type, four different wave conditions were tested. These were specifically chosen to ensure a range of overtopping magnitudes.

A number of wave conditions were initially tested, with three of those ($TS01$, $TS02$ and $TS05$) being deemed suitable for both of the structural geometries and were therefore tested with both structures in position. Due to the differing wave–structure interaction between the two different types of structure, the fourth wave condition for each case needed to be different. For the smooth slope, a small wave condition resulting in low overtopping ($TS03$) was chosen; however, this resulted in no overtopping of the vertical wall structure, and therefore it could not be used for this structure. Instead, a further wave condition was chosen for the vertical structure only ($TS07$). This resulted in enough wave overtopping to allow statistical analysis of the variation.

A summary of the incident target wave conditions, used in these tests, prescribed at the paddle are shown in Table 1. Here, $H_{m0,p}$ is the spectral significant wave height at the paddle, $T_{m-1,0}$ is the mean spectral period which can be defined as $T_{m-1,0} = \frac{m_{-1}}{m_0}$, where m_{-1} and m_0 are the spectral moments of order for -1 and 0, respectively. T_p is the peak period, d_p is the water depth at the paddle and R_c is the freeboard, or crest height above the still water level (SWL) with $R_c/H_{m0,p}$ being the relative freeboard, also known as R^*. These target values of R_c/H_{m0} were chosen to cover a range of overtopping from moderate to high.

Table 1. Incident wave conditions for the JONSWAP spectra random wave laboratory tests.

Test	Paddle							Toe					
	$H_{m0,p}$	$T_{m-1,0}$	T_p	d_p	$\dfrac{H_{m0,p}}{d_p}$	$s_{m-1,0}$	$\dfrac{R_c}{H_{m0,p}}$	H_{m0}	$T_{m-1,0}$	R_c	d_t	$\dfrac{H_{m0}}{d_t}$	$\dfrac{R_c}{H_{m0}}$
	(m)	(s)	(s)	(-)	(-)	(-)	(-)	(m)	(s)	(m)	(m)	(-)	(-)
TS01-SS	0.06	0.92	1.01	0.22	0.27	0.045	1.00	0.043	0.98	0.06	0.09	0.48	1.40
TS05-SS	0.05	0.85	0.93	0.22	0.22	0.044	1.20	0.038	0.89	0.06	0.09	0.42	1.58
TS02-SS	0.04	0.78	0.86	0.22	0.18	0.042	1.50	0.032	0.78	0.06	0.09	0.36	1.88
TS03-SS	0.03	0.64	0.70	0.22	0.14	0.047	2.00	0.020	0.77	0.06	0.09	0.22	3.00
TS01-VW	0.06	0.92	1.01	0.22	0.27	0.045	1.00	0.043	0.98	0.06	0.09	0.48	1.40
TS07-VW	0.05	1.13	1.24	0.22	0.22	0.025	1.20	0.040	1.33	0.06	0.09	0.44	1.50
TS05-VW	0.05	0.85	0.93	0.22	0.22	0.044	1.20	0.038	0.89	0.06	0.09	0.42	1.58
TS02-VW	0.04	0.78	0.86	0.22	0.18	0.042	1.50	0.032	0.78	0.06	0.09	0.36	1.88

In addition, Table 1 shows a summary of incident wave conditions measured at the location of the structure toes. These were obtained by running each wave condition in the flume without the structures in place, so that there was no reflection from the structures. During these tests, the foreshore remained in the flume, and a porous layer was placed at the end of the flume to absorb the waves. In Table 1, H_{m0}/d_t is the local wave height to local water depth ratio, where d_t is the water depth at the structure toe.

All tests were carried out using 1000 randomly generated irregular waves. This number was chosen as it is recommended in [1] as statistically representative of a sea state guaranteeing consistent results. Note that, in the following, we use the acronyms SS for the sloped structure and VW for the vertical wall when referring to the specific test, e.g., TS01-SS indicates the test on the sloped structure using the incident wave spectrum TS01.

3.2. Repetitions with Non-Random and Random Seeding

For each test condition, the initial seeding value was randomly altered to produce a different individual wave time series. One-hundred unique time series were generated for each test condition; this number of tests was previously found to be suitable for establishing the variability in q in numerical modelling [25]. These 100 initial seeding values were generated using a random number generator, and were different across each experimental condition. This means that, in total, 800 individual physical model tests were conducted to examine the effect of this random seeding.

In addition, since it has been suggested that repeatability of two nominally identical flume experiments may only be within 25% [24], it was also useful to quantify this variability separately to that introduced by the random seeding; it was therefore important to carry out a number of experiments using an identical initial seeding value, and therefore wave time series. This resulted in a supplementary 20 identical runs carried out for each wave condition and structure, resulting in an additional 160 tests.

3.3. Still Water Level Initial Conditions

During the preliminary experiments, it was found that the control of the initial SWL was a significant factor in determining the variability of the subsequent overtopping. Due to the small scale used in these experiments, even tiny variations in this water level at the beginning of each test can effect the quantity of overtopping. In addition, during these experiments, there was also a small loss of water within the flume over a period of time, making it difficult to maintain a constant SWL at the start of each test, so the following methodology was developed to minimise this problem and was used in all tests carried out.

A manual point gauge was placed on top of the wave flume to establish an initial SWL. This consisted of a small pointed metal rod which could be manually adjusted to precisely touch the water surface. A measurement reading was then taken using the attached graduated scale. This measurement, in relation to the top of the flume wall, was recorded to ensure that subsequent tests produced the same reading when measuring the initial water surface.

At the start of each test, if the point gauge was not touching the water surface, then water would be added to the flume, until this was achieved. At this point, the three wave gauges were used to measure the water surface for approximately 20 s, establishing the initial SWL for that particular set of tests.

A test was then carried out with an identical time series and wave conditions as a test carried out previously. The overtopping measurements from this new test were then compared with those obtained previously. If the new test produced an overtopping depth measured in the tank using the wave gauge within 0.5 mm of the one previously measured, then it was concluded that the initial water level in the flume was accurate.

At the end of each experimental run, the point gauge was again placed on the flume, and the water that had entered the overtopping tank during the test would be returned to the main flume, until the SWL met the point gauge. Once the water had settled down, the electronic wave gauges were used to measure the water surface again. These results were then compared with the original measurements at the start of that particular testing regime. If the results produced a variation in the water level of more than 0.5 mm, then the test could not be used and water was added to the flume until a variation of less than 0.5 mm was achieved and the test repeated.

3.4. Wave Generation

To ensure the accuracy of the experimental results, it is important to minimise any laboratory effects; this includes the presence of nonlinear effects caused by the mechanical method of wave generation. To investigate this issue, the wave height distributions both in front of the paddle and at the toe of the structure are considered for one of the wave conditions, namely *TS01*. The two distributions have been plotted in Figure 3 and compared with the expected deep water Rayleigh distribution as identified in the work of [30]. In this figure, the individual wave heights (H) have been normalised with the mean wave height (H_m).

The waves near the paddle closely match the Rayleigh distribution indicating that the paddle is indeed accurately reproducing a suitable sea state. The distribution at the structure toe shows a divergence from the Rayleigh distributions; this is caused by shoaling, triad interactions and depth-induced wave breaking. This agrees with the work of [31] which demonstrated a similar downward curved relation for the taller waves in shallow water like that present at the structure toe in this work. Further work showed these results to be consistent throughout the different wave conditions, and this additional analysis can be found in [32].

In addition, it was also important to check that the waves being generated matched the desired JONSWAP spectrum. The measure wave time series for each test was analysed in the frequency domain, and has been plotted in Figure 4, where f is the frequency and $S(f)$ is the energy density function.

This figure shows that the offshore wave energy density closely matches the shape of the target JONSWAP spectrum for this wave condition, suggesting the paddle is accurately reproducing this spectrum. Like the wave height distribution, the spectrum transforms closer to the structure toe due to shoaling, wave breaking, and reflection from the structure itself. Similarly to the wave height distribution, further work showed these results to be consistent throughout the different wave conditions, and this additional analysis can be found in [32].

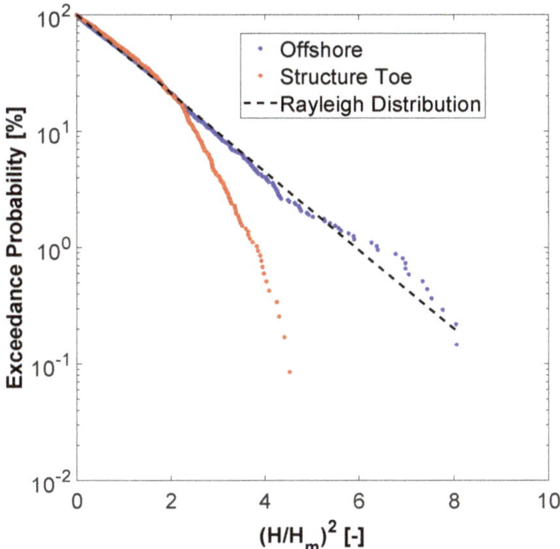

Figure 3. Distribution of the measured incident wave heights for *TS01* conditions generated by the wave paddle with comparison against expected Rayleigh Distribution.

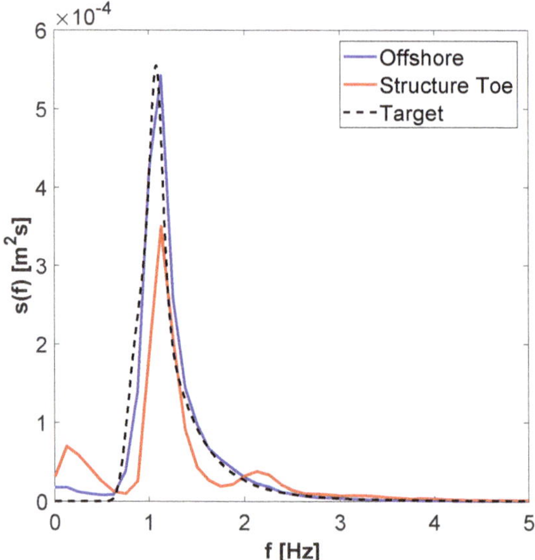

Figure 4. Measured spectra at different locations in the wave flume for *TS01* conditions generated by the wave paddle.

4. Results

4.1. Overtopping Discharge

First, we consider the variation in the overtopping discharge (q), for the 20 tests which were repeated using the exact same time series. The overtopping discharge was calculated for each test, with the results plotted using the dimensionless parameters, Q^* and R^* in Figure 5, defined as $Q^* = \frac{q}{\sqrt{gH_{m0}^3}}$ and $R^* = R_c/H_{m0}$. Using the dimensionless values, it allows a direct comparison between the different tests.

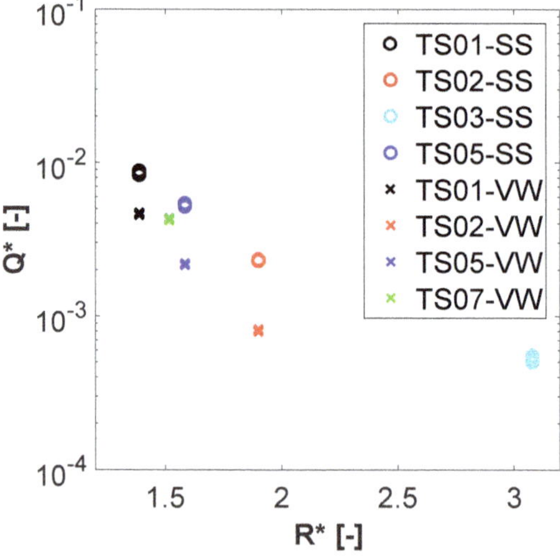

Figure 5. Graph showing R^* plotted against Q^* for each test conditions with identical offshore wave time series.

Figure 5 shows the results from both the smooth slope, and the vertical wall tests. It can be observed that there is some variation present in these nominally identical test runs. The magnitude of this variation remains fairly consistent across both the different levels of overtopping, and both structure types. To quantify this further, the variation in Q^* is calculated as a percentage difference for each test condition using the following formula:

$$Diff_{\%} = (1 - \frac{X_{min}}{X_{max}}) \times 100, \tag{1}$$

where X_{min} is the minimum value of the parameter of interest measured and X_{max} is the maximum value measured. The results of this are shown in Table 2.

Table 2. Percentage difference in H_{m0} and Q^* for non-random seeding tests.

Test	Q^* $Diff_{\%}$	H_{m0} $Diff_{\%}$
TS01-SS	6.13	1.73
TS02-SS	2.01	3.23
TS03-SS	10.10	5.87
TS05-SS	5.32	1.38
TS01-VW	4.61	1.42
TS02-VW	4.40	1.23
TS05-VW	4.19	1.06
TS07-VW	4.54	1.26

The results observed here experience a level of variation that is significantly lower than the 25% suggested in the work of [24] for both of the structure types, and for the different levels of overtopping. At this small scale, this level of variation is the equivalent to differences of less than a few millimetres in the depth detected in the tank, suggesting that at larger scales this variation would decrease further. The percentage difference in H_{m0} is also calculated here (see Table 2) to see if the variation measured is due to slight differences in the generated wave time series. The values for the percentage difference in H_{m0} were generally found to be less than those for the Q^* suggesting that the variation in Q^* does not purely originate from variation in H_{m0} but other laboratory effects.

Next, we examine the variation in the overtopping discharges based on the random seeding tests. To compare the variation in the prediction of overtopping from the physical model with that obtained when using empirical methods, the results were also compared with those calculated using formulae found within the latest version of [1]. Firstly, we examined the smooth slope structure, the most suitable formula for the conditions presented here has been developed from the work of [10]. The formula used is detailed below, or can be found as Equation 5.18 in the latest version of [1], and calculates the predicted dimensionless overtopping for the structure:

$$Q^* = \frac{q}{\sqrt{gH_{m0}^3}} = a \exp[-(b\frac{R_c}{H_{m0}})^c], \tag{2}$$

where $a = 0.09 - 0.01(2 - \cot \alpha)2.1$ for $\cot \alpha < 2$ and $a = 0.09$ for $\cot \alpha \geq 2$, $b = 1.5 + 0.42(2 - \cot \alpha)1.5$, with a maximum of $b = 2.35$ and $b = 1.5$ for $\cot \alpha \geq 2$ and $c = 1.3$.

This equation gives the mean value approach for wave overtopping and for the design or safety assessment approach; one standard deviation should be added providing the confidence bounds of the formulae. For the two coefficients a and b, this means that one should take $(1 + \sigma'(a))a = 1.15a$ and $(1 - \sigma'(b))b = 0.9b$, for a and b, where $\sigma' = \sigma/\mu$ with μ being the mean and σ is the standard deviation, which is calculated as $\sigma = \sqrt{\frac{1}{N-1} \sum_{i=1}^{N}(x_i - \mu)^2}$ where N is the number of samples considered.

Results of the comparison are plotted in Figure 6.

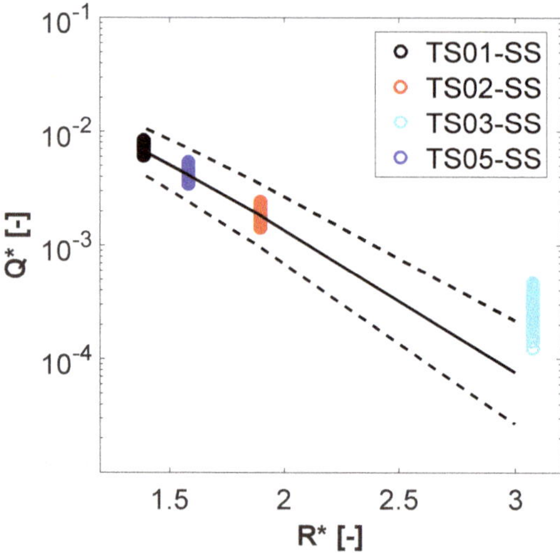

Figure 6. Graph showing R^* plotted against Q^* for random seeding test with smooth slope structure. Solid line: Empirical prediction. Dashed line: Confidence Bounds.

It can be observed here that the results from the physical model show very good agreement with those obtained using the empirical method, in all but the very lowest overtopping condition. The mismatch at the lower end of the results is most likely due to the calibration of the equation based on the work of [10]. This was based on structures with small relative freeboards, which is not the case for the lowest level of overtopping. One thing that can clearly be observed in these results is that, although there is variation in the prediction of overtopping in the physical model, it is well within the confidence bounds of the empirical method.

To compare the results from the vertical wall tests with the empirical methods, we must use a different formula. As no significant breaking was occurring in front of the vertical wall in the tests, we consider it to be in relatively deep water, where waves are not being significantly influenced by the sloping foreshore, we therefore used Equation 7.1 from [1], which is shown below:

$$Q^* = \frac{q}{\sqrt{gH_{m0}^3}} = 0.0047 \exp\left[-2.35\left(\frac{R_c}{H_{m0}}\right)^{1.3}\right]. \tag{3}$$

The results from the vertical wall tests using the physical model and those calculated using the empirical prediction are plotted in Figure 7.

Similarly to the smooth slope tests, the results show good agreement with the empirical prediction, this time across all of the tests carried out. It is also clear that the variation in the measurement of overtopping using the physical model is still within the confidence bounds used for the empirical method. It can also be observed across both sets of tests that the variation in overtopping is clearly dependent on the level of overtopping that occurred, with variation being highest in lower overtopping conditions, generally found when the relative freeboard is larger.

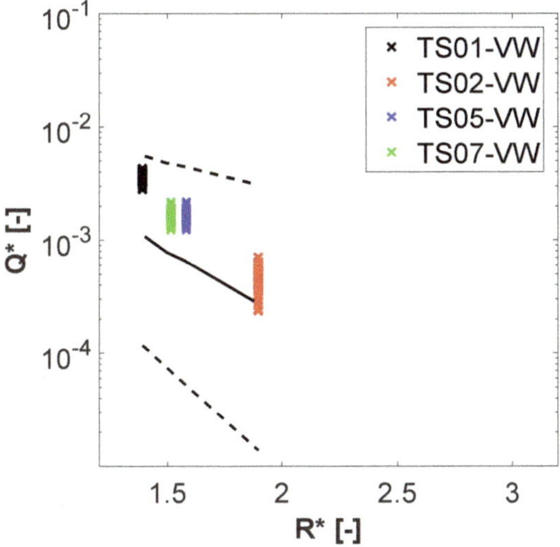

Figure 7. Graph showing R^* plotted against Q^* for random seeding test with vertical wall structure. Solid line: Empirical prediction. Dashed line: Confidence Bounds.

Using the same method as for the non-random seeding tests, the percentage differences in Q^* have been calculated and can be seen in Table 3. The percentage difference calculated using the empirical formulae is also included in this table. There are three main points that should be acknowledged from these results. Firstly, it can clearly be observed here that, unlike in the non-random seeding tests, the variation in Q^* is highly dependent on the magnitude of overtopping occurring. The tests with smaller incoming wave conditions, or those based on the vertical wall, and therefore experienced lower levels of overtopping show larger variation than those that experienced higher levels of overtopping. The second point is that the magnitude of the variation, in terms of $Diff_\%$ for the low overtopping tests, is shown to be up to 73.37%. This is a significant difference in the prediction of the overtopping discharge from the same wave energy spectrum, suggesting that overtopping could be significantly under-predicted depending on the exact time series utilised. The third point is that, although relatively large variation is observed across the physical model tests, these are still a lot less than those incorporated via the confidence bounds into the empirical methods.

Additionally, the percentage difference in H_{m0} has also been calculated, and can be found in Table 3. Although this variation shows slightly more variability than those for the non-random seeding tests, it is still significantly smaller than that found for Q^*. This suggests that some of the wave sequences produced by the paddle may not fit the wave statistics as closely as others, but does not significantly change the findings with regard to variability in Q^*.

Table 3. Percentage difference in H_{m0} and Q^* for random seeding tests.

Test	Experimental Q^* $Diff_\%$	Empirical Q^* $Diff_\%$	H_{m0} $Diff_\%$
TS01-SS	28.00	61.33	8.29
TS02-SS	40.65	71.75	13.32
TS03-SS	73.37	87.61	9.84
TS05-SS	35.72	65.45	9.78
TS01-VW	35.21	97.88	8.86
TS02-VW	66.86	99.56	11.20
TS05-VW	42.89	98.83	10.40
TS07-VW	39.82	98.56	10.49

4.2. Probability of Overtopping

It has been observed here that the variation in the overtopping discharge is related to the level of overtopping; therefore, analysis was carried out to quantify further whether the magnitude of variation in the discharge is directly related to the probability of overtopping for the physical model, as found in the analysis of numerical modelling prediction of overtopping in [25], which focussed solely on smooth slope structures.

The results from the physical model, with both types of structure, can be seen in Figure 8, where the probability of overtopping (P_{ov}) is plotted against the overtopping discharge (q). P_{ov} is calculated by $\left(\frac{N_{ow}}{N_w}\right) \times 100$, where N_{ow} is the number of waves that overtopped the structure and N_w is the total number of waves generated by the wave paddle in each test. N_{ow} was calculated using the peak detection method described earlier in this work, in order to identify each overtopping event, providing an approximation of the total number of events that occurred. N_w was calculated by applying the zero-crossing method to the incoming waves measured at $WG2$ at the toe of the foreshore, identifying each individual wave, and therefore providing a value for the total number of waves in the test.

These results show that the different types of structures appear to have no direct impact on the magnitude of the variation in q. This means that, where the magnitude of overtopping is similar, the level of variability is similar, irregardless of the shape of the structure or the incoming wave parameters, suggesting that P_{ov} is indeed the main influencing factor controlling variability.

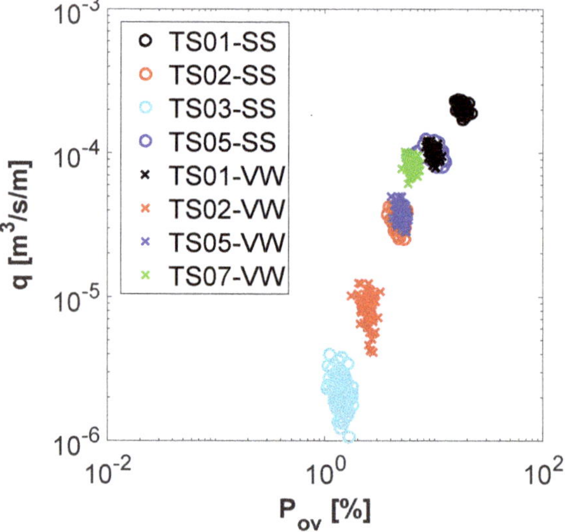

Figure 8. Graph showing P_{ov} plotted against q for random seeding test with both structures.

4.3. Individual and Maximum Overtopping Volumes

Overtopping discharge and probability of overtopping are useful metrics in the design of coastal structures, but in terms of the safety of users and potential damage, the individual overtopping volumes are also important. There are two characteristics to be examined here, both the distributions of the individual overtopping volumes V_{ov}, and the maximum overtopping volume V_{max}. As we have already found that variability is highly dependent on P_{ov}, the V_{max} is considered with relation to this parameter and is plotted in Figure 9.

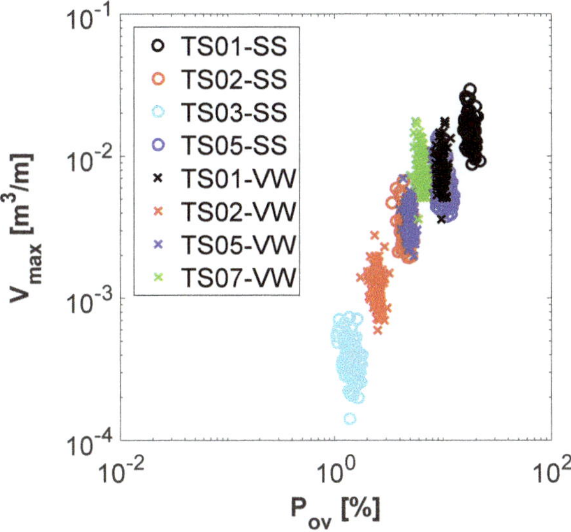

Figure 9. Graph showing P_{ov} plotted against V_{max} for a random seeding test with both structures.

Similarly to q, it can be seen that V_{max} does show variation, but still only within one order of magnitude. There is a small increase in variation which is most likely due to the extreme nature of the V_{max} value, as opposed to the average value of q, i.e., a single large volume has more influence when not averaged across all of the volumes. If we look at the variation in P_{ov}, it can also be seen that the value of P_{ov} has less influence on this variability with all the tests showing a similar level of variation for all the levels of overtopping.

To examine this further, the individual volume distributions are considered. Firstly, the empirical cumulative distribution functions of the individual overtopping volumes for each of the 100 random seeding tests for each wave condition with the smooth slope have been plotted in Figure 10. These have been normalised by dividing each individual volume measured, by the mean of the volumes measured for each test; this allows a direct comparison between the different levels of overtopping.

These graphs clearly show the increase in variation of these distributions of individual overtopping volumes due to the overall decrease in overtopping. The number of overtopping events can also be seen to decrease as the overall overtopping magnitude decreases, thus increasing the significance of each individual overtopping volumes in each time series.

The probability distribution of V_{ov} was investigated by a number of researchers ([10,11,33]) and found to generally follow a two parameter Weibull distribution:

$$P_v = P(V_{ov} \leq V_{ov}) = 1 - exp\left[-\left(\frac{V_{ov}}{a}\right)^b\right], \tag{4}$$

where P_v is the exceedance probability of each overtopping volume. The scale factor, a, and the shape factor, b, can vary depending on the exact empirical method chosen which is selected based on the conditions present. In this work, the distributions **were** fitted with an idealised Weibull distribution. These have been obtained by fitting the Weibull distribution to each individual time series, and then using the mean values calculated for the scale and shape parameters, along with the extreme values for the outer limits. As expected, the Weibull distribution shows a good match with the measured distributions across all of the levels of overtopping, albeit with differing values for shape and scale parameters.

Additionally, on these graphs, two distributions from the measured non-random seeding tests have been included. These represent the extremes of the measurements of V_{ov} for these tests. As expected from the earlier analysis, the variation in these is significantly lower than when random seeding is introduced. This is particularly noticeable in the lower overtopping tests, where the variation magnitude differs more.

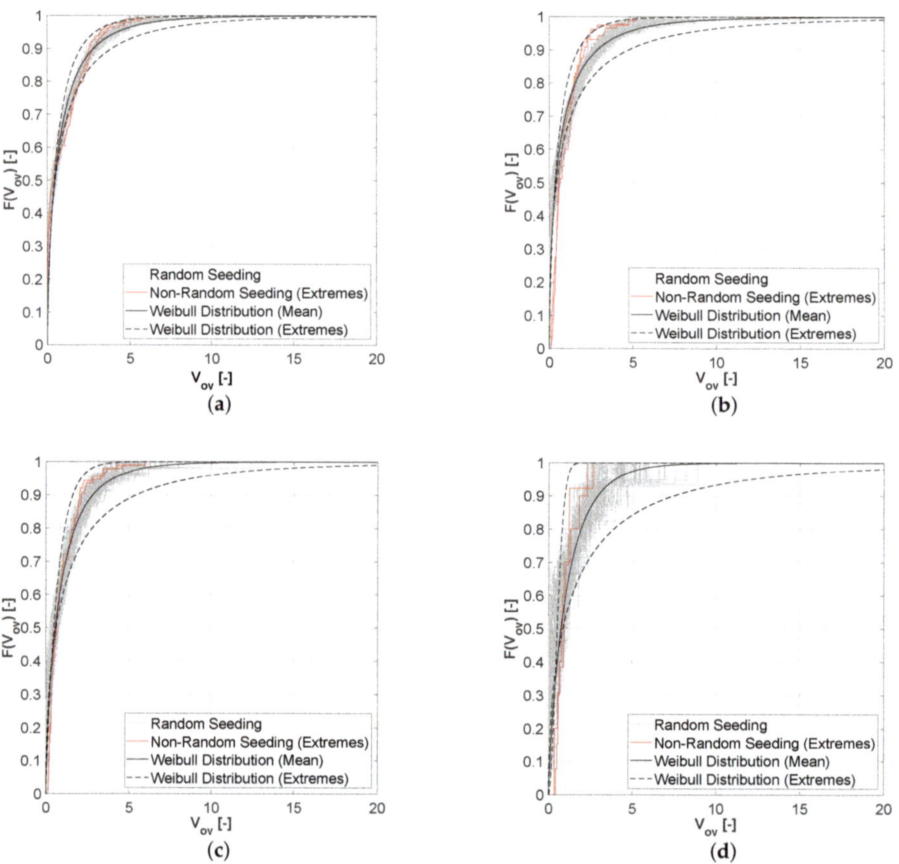

Figure 10. Cumulative density functions showing distributions of individual overtopping volumes for smooth slope. (**a**) TS01-SS; (**b**) TS05-SS; (**c**) TS02-SS; (**d**) TS03-SS.

Now, the individual volumes for the vertical wall are considered using the same method, resulting in Figure 11.

These graphs clearly show the same behaviour as those using the smooth slope structure; again, it can be seen that the variation increases with the decrease in overall overtopping magnitude. It has also been shown once again that this is not directly influenced by the type of structure, confirming that P_{ov} is the main parameter influencing the variability in overtopping prediction due to the random seeding of the offshore wave time series.

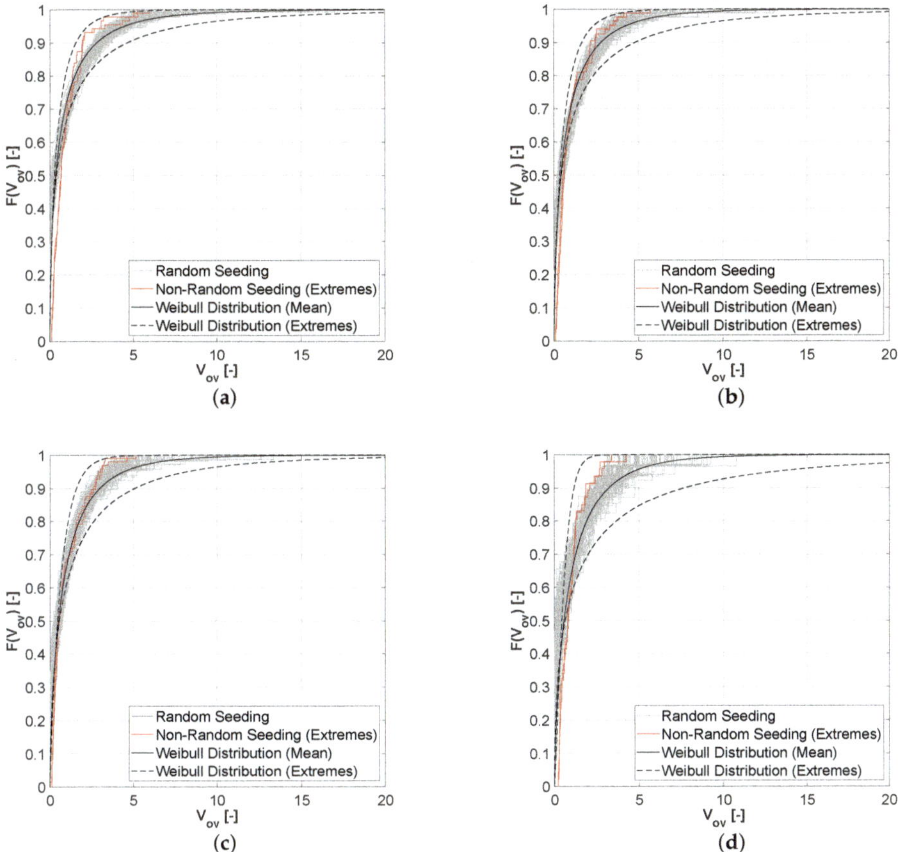

Figure 11. Cumulative density functions showing distributions of individual overtopping volumes of vertical wall.(**a**) TS01-VW; (**b**) TS07-VW; (**c**) TS05-VW; (**d**) TS02-VW.

5. Discussion

The variability in the physical model in predicting q due to both different seeding and laboratory effects has been established. It is important to compare these two sets of results, so that the variabilities can be separated. This is carried out by calculating the relative standard deviation (σ') in q for each of the tests. σ' represents a standardised measure for the dispersion of a frequency distribution, and is calculated as the ratio of the standard deviation (σ) to the mean (μ) of the results ($\sigma' = \frac{\sigma}{\mu}$), in this case using the overtopping discharge. These are plotted against the relative freeboard in Figure 12 for both structures.

It can be seen on these graphs that σ' for the non-random seeding remains fairly consistent throughout the different tests across both structural geometries, whereas the σ' for the random seeding tests increases significantly as R^* increases. There is a larger increase in σ' observed at a lower value of R^* for the vertical wall; this is due to the lower overtopping discharges, and subsequent greater variability experienced by the vertical wall than the smooth slope for the same wave conditions and crest height.

Across the two tested geometries, the value of σ' for the non-random seeding equates to approximately 1%. This suggests that, within the physical modelling of overtopping, there is always a very small amount of variation that can be equated simply to laboratory effects, and cannot easily be removed.

When the variation introduced due to random seeding is considered, results show that the magnitude of this is influenced by the magnitude of overtopping, i.e., the less overtopping, the higher the variation. This ranges from 6–25% in this work, and suggests that the small variation introduced due to laboratory effects is more significant in the tests with greater overtopping than those with small overtopping levels where the larger variation lowers the significance of this in the overall variability of the results.

In this work, a total of 100 random seeding tests have been carried out for each test condition. This is a large number of tests, and, in reality, it is not feasible to carry out this many repetitions; it was therefore decided to establish a more suitable value for the number of repetitions required. This was carried out by taking the full overtopping discharge calculated for each of the 100 tests and randomly sampling with different number of observations, namely 2, 5, 10, 20 and 50 samples. For each of these subsets of results, the values of σ' were calculated, and analysed to see when the values of σ' converged, suggesting that the minimum number of tests required had been met. When $P_{ov} < 10\%$, it was found that a minimum of 20 repetitions was required; this could decrease to 10 repetitions when $P_{ov} > 10\%$. This agrees with the findings of [28], which found that, when $Q^* < 10^{-3}$, at least 20 tests provided a good estimate; this is the region where $P_{ov} < 10\%$.

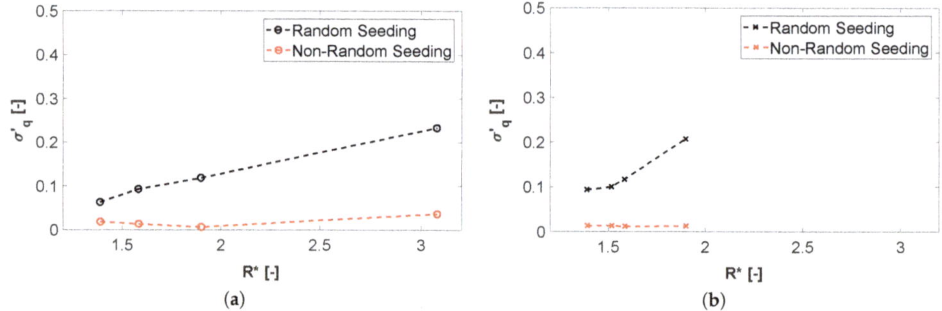

Figure 12. σ' of q for Random Seeding and Non-Random Seeding tests. (**a**) smooth slope; (**b**) vertical wall.

6. Conclusions

The paper quantifies the influence on wave overtopping of the initial random seeding on the generation of wave time series from wave energy spectra in a small scale physical model. Overall, the findings from these experiments agree with previous numerical investigations of the same phenomenon [25,26].

Although the experiments carried out in this work are on a small scale, the relative comparison of experiments within this dataset allows for quantifying the magnitude of the variability among Q^* and establishing a relationship with the probability of overtopping. The scale effects present in the experiment do not allow for extrapolate definitive conclusions valid at all scales but provide some guidelines for experiments in similar scales. Note that [14] reported that the majority of tests in the new wave overtopping database comes from tests with $0.05 < H_{m0} < 0.15\,m$.

The results have also been compared with the most up-to-date formulae found in [1]. The variability found in the present results is lower than the overall variability in the formulae, with all results falling within the provided confidence bounds unless the conditions deviated from the exact criteria of the formulae used. However, the uncertainty introduced by iso-energetic sequences of waves is still significant. By using $Diff_{\%}$, it is found that this variability is approximately between 20% and 75% of the Q^*_{max}, i.e., for $R^* > 2$, the variability could be comparable with Q^*_{max}. This indicates the need of taking this factor into account when experimental tests are performed, by ideally repeating tests in this region of R^* to estimate the range of Q^*. The number of repetitions required has been found to be a minimum of 20 when $P_{ov} < 10\%$, but this can decrease to 10 repetitions when P_{ov} is higher.

Author Contributions: This work was carried out as part of the PhD of H.E.W.; H.E.W. and R.B. planned the test campaign for this paper. The experimental tests and analysis were performed by H.E.W., with additional analysis from R.B. The original draft of the paper was written by H.E.W.; R.B., A.R. and N.D. performed a detailed review and editing of the draft paper and contributed with valuable suggestions. Supervision of the project was by R.B. and N.D. Funding acquisition was by R.B.

Funding: Hannah Williams and Riccardo Briganti were supported by the Engineering and Physical Sciences Research Council Career Acceleration Fellowship (EP/I004505/1).

Acknowledgments: The authors would also like to thank Steve Gange and Marco Matteucci at the University of Nottingham for their support during the experimental tests carried out in this work.

Conflicts of Interest: The authors declare no conflict of interest.

References

1. Van der Meer, J.W.; Allsop, N.W.H.; Bruce, T.; De Rouck, J.; Kortenhaus, A.; Pullen, T.; Schüttrumpf, H.; Troch, P.; Zanuttigh, B. *EurOtop 2018. Manual on Wave Overtopping of Sea Defences and Related Structures. An Overtopping Manual Largely Based on European Research, but for Worldwide Application*; Environment Agency: Bristol, UK, 2018.
2. Troch, P.; Geeraerts, J.; van de Walle, B.; De Rouck, J.; Van Damme, L.; Allsop, N.W.H.; Franco, L. Full-scale wave-overtopping measurements on the Zeebrugge rubble mound breakwater. *Coast. Eng.* **2004**. *51*, 609–628. [CrossRef]
3. Briganti, R.; Bellotti, G.; Franco, L.; De Rouck, J.; Geeraerts, J. Field measurements of wave overtopping at the rubble mound breakwater of Rome Ostia yacht harbour. *Coast. Eng.* **2005**, *52*, 1155–1174. [CrossRef]
4. Pullen, T.; Allsop, N.W.H.; Bruce, T.; Pearson, J. Field and laboratory measurements of mean overtopping discharges and spatial distributions at vertical seawalls. *Coast. Eng.* **2009**, *56*, 121–140. [CrossRef]
5. Franco, L.; Geeraerts, J.; Briganti, R.; Willems, M.; Bellotti, G.; De Rouck, J. Prototype measurements and small-scale model tests of wave overtopping at shallow rubble-mound breakwaters: The Ostia-Rome yacht harbour case. *Coast. Eng.* **2009**, *56*, 154–165. [CrossRef]
6. Pearson, J.; Bruce, T.; Allsop, N.W.H. Prediction of wave overtopping at steep seawalls—Variabilies and uncertainties. In Proceedings of the Fourth International Symposium on Ocean Wave Measurement and Analysis, San Francisco, CA, USA, 2–6 September 2001; Volume 1, pp. 1797–1808.
7. Van Gent, M.R.A.; van den Boogaard, H.F.P.; Pozueta, B.; Medina, J.R. Neural network modelling of wave overtopping at coastal structures. *Coast. Eng.* **2007**, *54*, 586–593. [CrossRef]
8. Reis, M.T.; Neves, M.G.; Hedges, T. Investigating the Lengths of Scale Model Tests to Determine Mean Wave Overtopping Discharges. *Coast. Eng. J.* **2008**, *50*, 441–462. [CrossRef]
9. Bruce, T.; van der Meer, J.W.; Franco, L.; Pearson, J.M. Overtopping performance of different armour units for rubble mound breakwaters. *Coast. Eng.* **2009**, *56*, 166–179. [CrossRef]
10. Victor, L.; van der Meer, J.W.; Troch, P. Probability distribution of individual wave overtopping volumes for smooth impermeable steep slopes with low crest freeboards. *Coast. Eng.* **2012**, *64*, 87–101. [CrossRef]
11. Nørgaard, J.Q.H.; Lykke Andersen, T.; Burcharth, H.F. Distribution of individual wave overtopping volumes in shallow water wave conditions. *Coast. Eng.* **2014**, *83*, 15–23. [CrossRef]
12. Van Doorslaer, K.; Romano, A.; De Rouck, J.; Kortenhaus, A. Impacts on a storm wall caused by non-breaking waves overtopping a smooth dike slope. *Coast. Eng.* **2017**, *120*, 93–111. [CrossRef]
13. Martinelli, L.; Ruol, P.; Volpato, M.; Favaretto, C.; Castellino, M.; De Girolamo, P.; Franco, L.; Romano, A.; Sammarco, P. Experimental investigation on non-breaking wave forces and overtopping at the recurved parapets of vertical breakwaters. *Coast. Eng.* **2018**, *141*, 52–67. [CrossRef]
14. Van der Meer, J.W.; Verhaeghe, H.; Steendam, G.J. The new wave overtopping database for coastal structures. *Coast. Eng.* **2009**, *56*, 108–120. [CrossRef]
15. Dodd, N. A numerical model of wave run-up, overtopping and regeneration. *J. Waterw. Port Coast. Ocean Eng.* **1998**, *124*, 73–81. [CrossRef]
16. Hubbard, M.E; Dodd, N. A 2-d numerical model of wave run-up and overtopping. *Coast. Eng.* **2002**, *47*, 1–26. [CrossRef]

17. Reeve, D.E.; Soliman, A.; Lin, P.Z. Numerical study of combined overflow and wave overtopping over a smooth impremeable seawall. *Coast. Eng.* **2008**, *55*, 155–166. [CrossRef]
18. Ingram, D.M.; Gao, F.; Causon, D.M; Mingham, C.G.; Troch, P. Numerical investigations of wave overtopping at coastal structures. *Coast. Eng.* **2009**, *56*, 190–202. [CrossRef]
19. Tonelli, M.; Petti, M. Numerical simulation of wave overtopping at coastal dukes and low-crested structures by means of a shock-capturing Boussinesq model. *Coast. Eng.* **2013**, *79*, 75–88. [CrossRef]
20. Suzuki, T.; Altomare, C.; Veale, W.; Verwaest, T.; Trouw, K.; Troch, P.; Zijlema, M. Efficent and robust wave overtopping estimation for impermeable coastal structures in shallow foreshores using SWASH. *Coast. Eng.* **2017**, *122*, 108–123. [CrossRef]
21. Akbari, H. Simulation of wave overtopping using an improved SPH method. *Coast. Eng.* **2017**, *126*, 51–68. [CrossRef]
22. Castellino, M.; Sammarco, P.; Romano, A.; Martinelli, L.; Ruol, P.; Franco, L.; De Girolamo, P. Large impulsive forces on recurved parapets under non-breaking waves. A numerical study. .*Coast. Eng.* **2018**, *136*, 1–15. [CrossRef]
23. McCabe, M.V.; Stansby, P.K.; Apsley, D.D. Coupled wave action and shallow-water modelling for random wave runup on a slope. *J. Hydraul. Res.* **2011**, *49*, 512–522. [CrossRef]
24. McCabe, M.V.; Stansby, P.K.; Apsley, D.D. Random wave runup and overtopping a steep sea wall: Shallow-water and Boussinesq modelling with generalised breaking and wall impact algorithms validated against laboratory and field measurements. *Coast. Eng.* **2013**, *74*, 33–49. [CrossRef]
25. Williams, H.E.; Briganti, R.; Pullen, T. The role of offshore boundary conditions in the uncertainty of numerical prediction of wave overtopping using nonlinear shallow water Equations. *Coast. Eng.* **2014**, *89*, 30–44. [CrossRef]
26. Williams, H.E.; Briganti, R.; Pullen, T.; Dodd, N. The Uncertainty in the Prediction of the Distribution of Individual Wave Overtopping Volumes Using a Nonlinear Shallow Water Equation Solver. *J. Coast. Res.* **2016**, *32*, 946–953. [CrossRef]
27. Palemón-Arcos, L.; Torres-Freyermuth, A.; Pedrozo-Acuña, A.; Salles, P. On the role of uncertainty for the study of wave–structure interaction. *Coast. Eng.* **2015**, *106*, 32–41. [CrossRef]
28. Romano, A.; Bellotti, G.; Briganti, R.; Franco, L. Uncertainties in the physical modelling of the wave overtopping over a rubble mound breakwater: The role of the seeding number and of the test duration. *Coast. Eng.* **2015**, *103*, 15–21. [CrossRef]
29. Hughes, S. *Physical Models and Laboratory Techniques in Coastal Engineering*; World Scientific: New York, NY, USA, 1993.
30. Longuet-Higgins, M.S. On the statistical distribution of the heights of sea waves. *J. Mar. Res.* **1952**, *9*, 245–266.
31. Battjes, J.A.; Groenendijk, H.W. Wave height distributions on shallow foreshores. *Coast. Eng.* **2000**, *40*, 161–182. [CrossRef]
32. Williams, H.E. Uncertainty in the Prediction of Overtopping Parameters in Numerical and Physical Models due to Offshore Spectral Boundary Conditions. Ph.D. Thesis, University of Nottingham, Nottingham, UK, 2015.
33. Van der Meer, J.; Janssen, J. *Wave Run-Up and Wave Overtopping at Dikes and Revetments*; Delft Hydraulics: Delft, The Netherlands, 1994.

Journal of
*Marine Science
and Engineering*

Article

Effects of Wave Orbital Velocity Parameterization on Nearshore Sediment Transport and Decadal Morphodynamics

Marcio Boechat Albernaz [1,*], Gerben Ruessink [1], H. R. A. (Bert) Jagers [2] and Maarten G. Kleinhans [1]

[1]   Department of Physical Geography, Faculty of Geosciences, Utrecht University, 3508TC Utrecht, The Netherlands; B.G.Ruessink@uu.nl (G.R.); M.G.Kleinhans@uu.nl (M.G.K.)
[2]   Deltares, 2600MH Delft, The Netherlands; Bert.Jagers@deltares.nl
*   Correspondence: m.boechatalbernaz@uu.nl

Received: 24 May 2019; Accepted: 15 June 2019; Published: 19 June 2019

Simple Summary: Sandy coasts evolve as a result of sand transport by waves and tides. Wave-generated flows near the seabed stir the sand into the water column, which can subsequently be transported in cross-shore and alongshore directions. As waves move shoreward into shallower depth, the shape of the near-bed flow changes, which affects the magnitude of sand transport. Until now, computer simulations of decadal coastal development did not contain this shape change adequately and, as a consequence, modelled coasts were predicted to erode or accrete too rapidly. Therefore, we implemented a novel wave shape module based on field data and tested its performance in predicting the decadal evolution of typical, well-monitored sites on sandy Dutch and US East coasts. The modified model now predicts realistic cross-shore profile evolution at both sites without excessive shoreline erosion or accretion. Also, the sand transport rates along the coast are better represented. This opens up the possibility to realistically model coastal evolution on the timescale of decades to a century.

Abstract: Nearshore morphological modelling is challenging due to complex feedback between hydrodynamics, sediment transport and morphology bridging scales from seconds to years. Such modelling is, however, needed to assess long-term effects of changing climates on coastal environments, for example. Due to computational efficiency, the sediment transport driven by currents and waves often requires a parameterization of wave orbital velocities. A frequently used parameterization of skewness-only was found to overfeed the coast unrealistically on a timescale of years—decades. To improve this, we implemented a recently developed parameterization accounting for skewness and asymmetry in a morphodynamic model (Delft3D). The objective was to compare the effects of parameterizations on long-term coastal morphodynamics. We performed simulations with default and calibrated sediment transport settings, for idealized coastlines, and compared the results with measured data from analogue natural systems. The skewness-asymmetry parameterization was found to predict overall stable coastlines within the measured envelope with wave-related calibration factors within a factor of 2. In contrast, the original parameterization required stronger calibration, which further affected the alongshore transport rates, and yet predicted erosion in deeper areas and unrealistic accretion near the shoreline. The skewness-asymmetry parameterization opens up the possibility of more realistic long-term morphological modelling of complex coastal systems.

Keywords: wave parameterization; wave skewness; morphodynamic modelling; sediment transport; delft3d; wave shape

1. Introduction

As deep-water linear waves approach shallow coastal zones, they begin to interact with the bottom and change their shape and orbital motion towards the shoreline. Along the propagation path, non-linearities arise with the waves first becoming skewed with a shorter, higher crest and longer shallower trough, and, in the shallow surf zone, gradually changing into asymmetric waves with a saw-tooth shape, pitched forward with a steep front and gentle rear [1]. This process modifies the near-bed orbital velocities, which impacts sediment transport and long-term, large-scale morphological evolution, here defined as years to decades covering kilometers. The wave shape transformation can be fully computed by means of phase-resolving models [2,3]; however, this requires large computational efforts, restricting simulations to hydrodynamics-only scenarios of small temporal–spatial scales (O hours-meters). Therefore, for reasons of computational efficiency, the wave shape transformation and orbital velocities are often parameterized in phase-averaged morphodynamic models. Nonetheless, the parameterization effects starting from the hydrodynamics of orbital motion up to long-term morphological evolution remain highly uncertain. Small deviations in orbital velocities between parameterizations combined with non-linear response of sediment transport makes morphological predictions rather uncertain and ultimately inaccurate over such long time-scales. Yet this is a societal relevant scale for coastal protection by sediment management in view of climate change and sea level rise [4–6].

Various parameterizations have been proposed and implemented in the past decades, for example, non-linear wave theories such as Second (or higher) Order Stokes [7,8], Stream Function [9] and the hybrid theory of Isobe and Horikawa [10] that combines 3rd Cnoidal and 5th Order Stokes. These and other theories are compared by Dean and Perlin [8]. In general, the parameterizations derived from scaled physical experiments reproduce only skewness (velocity skewness) and acknowledge that asymmetry (acceleration skewness) is not accounted for [7,10] and consequently predictions for shallower nearshore dynamics are rather inaccurate.

Cross-shore morphodynamic models of varying complexities account explicitly for physical processes coupled to sediment transport predictors in order to simulate nearshore dynamics. Wave non-linearities are incorporated through one of the aforementioned methods (e.g., [11,12]). Although cross-shore models have mainly been applied to hindcast nearshore sandbar behaviour in short time spans (O hours-days), a few attempts to simulate mild energy conditions on longer time scales have been reported [12,13]. Here, the onshore net sediment transport only occurred when wave non-linear orbital velocities were included, for example with a stream function [12–15] and with the skewness parameterization of Isobe and Horikawa [10,12,16]. In these models, the undertow and return currents are the main hydrodynamic contributors to offshore net sediment transport during storms and the short-wave non-linearity mainly contribute to the onshore transport during mild conditions [7,11,12]. Whereas the skewness-based models are adequate for simulating sand bar migration in the shoaling and outer surfzone, their performance in the inner surfzone and near the shoreline, where skewness decreases while asymmetry increases, was poor, mainly due to the overestimation of sand transport towards the shoreline [12,17]. Recently, a parameterization including skewness and asymmetry was derived from comprehensive field data [18]. The application of the Ruessink et al. [18] parameterization in a cross-shore model [17] demonstrated that neglecting the asymmetry led to poor performance. This could partly be compensated by enhancing skewness but resulted in unrealistically large deposition in the inner surfzone and shoreline displacement. On the other hand, the same model with skewness and asymmetry performed equally well for short-term surfzone sandbar migration without the unrealistic side-effects near the shoreline. Nevertheless, the performance of this parameterization has not yet been studied in the context of long-term morphological modelling.

Despite the acknowledged relevance in the cross-shore models, non-linearities of orbital velocities are not as well represented in process-based area morphodynamic models in depth-averaged (2DH) and three-dimensional (3D) configurations. These models are often intended to simulate complex coastal environments from deep water to the shoreline, including tidal inlets, tidal basins and estuaries that reproduce important features and dynamics between fluvial and coastal processes. By neglecting or oversimplifying the wave-orbital velocity shape, for example with linear wave theory [19–21] or skewness-only predictors [22–26], we jeopardize the reliability of predicting the evolution of these complex environments especially in the long-term. The morphodynamic feedback system of morphology, sediment transport and hydrodynamics is then essentially preset to evolve to non-realistic equilibrium situations. In addition to the wave parameterization, 2DH models lack vertical processes, such as undertow and have relatively coarser grids when compared to purely cross-shore models. Therefore, these vertically distributed hydrodynamic processes in the cross-shore direction need to be (over)compensated for this limitation, which is commonly done by reducing the onshore sediment transport to a certain factor, by means of sediment transport linear calibration factors, to balance the lack of offshore component [27,28].

By (over)calibrating the sediment transport, a mismatch may arise along depth strata and between cross-shore and alongshore sediment transport rates and associated morphological timescales. As a consequence, studies had to focus on either cross-shore (e.g., [25,28]) or alongshore processes (e.g., [24,27]), or prioritize the lower-mid or upper shoreface morphological performance [12,13,15]. Consequently, a model calibrated to match alongshore transport rates can wrongly predict shoreline retreat or progradation and vice-versa. This is an important limitation for model studies of mixed environments, such as ebb deltas, tidal inlets and estuaries wherein tidal bars and shoals form through interaction of combined fluvial, estuarine and coastal waves and currents. As a result, limited attempts have been made in simulating long-term morphology in these conditions. Most models either focused on long-term simulations without waves or only accounting for low energetic wave conditions aiming sediment stirring [29–34] or short-term simulations [21–24,26,27,35–37].

The hypothesis in this paper is that an important component of the overall nearshore sediment balance and morphological evolution in 2DH models derives from a more reliable parameterization of near-bed orbital velocities induced by wave skewness and asymmetry [15,17]. If implementation improves performance in long-term morphological modelling, then that would open up the possibilities to model mixed environments. Therefore, our aim is to compare long-term sediment transport and morphodynamic development in a 2DH area model with near-bed wave orbital velocities parameterized with a skewness-only method [10] with a skewness and asymmetry formulation [18]. We applied a comprehensive set of wave climates at two idealized coasts based on coastal sites in The Netherlands and USA, including locally generated short waves and swell conditions. Our assessment targets long-term nearshore profile equilibrium conditions in the cross-shore and the analogue response in the alongshore direction under varying parameterizations and sediment transport calibration scenarios. The morphological developments are compared with long-term measurements of cross-shore bed evolution.

2. Methods

To assess the morphodynamic performance of a skewness-only short-wave parameterization versus a skewness and asymmetry formulation, we implemented the Ruessink et al. [18] parameterization into Delft3D which already contains, as default, the skewness-only Isobe and Horikawa [10] formulation. Delft3D is an extensively applied morphodynamic model of finite differences, solving the momentum and continuity equations for unsteady shallow-water flow in depth-averaged or three-dimensional mode through the Navier-Stokes equation with hydrostatic pressure approximation [38]. The hydrodynamics are coupled with the SWAN (i.e., DELFT3D-WAVE) spectral wave model [39,40]. The equations from Ruessink et al. [18] (further referred to as RUE) were embedded into the source code (FLOW2D3D version 6.02.13.7658 from tag 7545 [41]) as an alternative to the currently operational Isobe and Horikawa [10] method (referred to as IH), modified by Grasmeijer [16] and van Rijn [42]. The near-bed orbital velocities are coupled with the TRANSPOR2004 (henceforth called VR04) sediment transport predictor [38,43]. We elected to use VR04 because it is well-calibrated on a wide range of environments computing current and wave-related sediment transport, including interaction between wave–current and intra-wave sediment transport and conceptually separates bed and suspended load for current and waves [43–46].

Below we describe the orbital velocity parameterization of IH and RUE, the sediment transport predictor, and the setup of the numerical models including the initial and boundary conditions and applied wave climate at Katwijk, The Netherlands and Duck, North Carolina, USA.

2.1. Parameterization of Wave Shape and Orbital Velocity

The near-bed orbital velocities are parameterized in IH and RUE based on the local root-mean-square wave height (H_{rms}), peak period (T_p) and water depth (h). The IH method after modifications from Grasmeijer [16], van Rijn [42] computes the wave shape and skewed orbital velocities from a hybrid wave theory combining a fifth-order Stokes and third-order cnoidal wave theory derived from laboratory experiments. The RUE method computes skewness and asymmetry derived from the Ursell number (Ur), and the resulting wave shape follows from Abreu et al. [47]. The RUE parameterization applies a functional fit to compute Skewness (Sk) and Asymmetry (As) based on extensive field measurements in contrast with previous methods, including IH, based solely on limited physical experiments. The IH method does not explicitly computes Sk and As, therefore we applied a simple skewness (R_u) and asymmetry (R_a) coefficients based on predicted peak velocity and acceleration values, respectively

$$R_u = \frac{u_{on}}{u_{on} + |u_{off}|}$$

and $\qquad\qquad\qquad\qquad\qquad\qquad\qquad\qquad\qquad\qquad\qquad$ (1)

$$R_a = \frac{a_{on}}{a_{on} + |a_{off}|} \quad .$$

Here, u_{on} is the peak onshore and u_{off} offshore orbital velocity in m/s; a_{on} is the maximum onshore and a_{off} offshore a_{off} acceleration in m/s^2. Coefficients larger than 0.5 represent deviations due to non-linearities. Below we review the basic concepts of each formulation while detailed equations are provided in Appendix A.

2.1.1. Isobe Horikawa [IH]

The original formulation of Isobe and Horikawa [10] parameterizes the wave shape and orbital velocities based on the offshore wave height and period, local water depth and bed slope without explicitly quantifying skewness. In order to be implemented in numerical models, this original formulation was adapted by Grasmeijer [16], also based on physical experiments, to compute the peak orbital velocities u_{on} and u_{off} with local wave height and period, but without the bed slope dependence. In Delft3D-VR04 the adapted version of Isobe and Horikawa [10], Grasmeijer [16] is implemented as default in order to calculate the intra-wave orbital velocity $u(t)$ within the wave period [42]. In summary, we can describe the steps of IH parameterization as follows. First, based on H_{rms}, T_p, h, IH introduces an empirically derived non-linear parameter for calculating the maximum onshore-directed velocity u_{on}; and the maximum offshore velocity u_{off} is indirectly calculated based on the amplitude velocity U_w. From the relative ratio of u_{on} and u_{off}, the onshore (T_{for}) and offshore period (T_{back}) are estimated. The velocity profile is then derived separately for onshore ($0 < t < T_{for}$) and offshore-directed ($T_{for} < t < T$) flows from the intra-wave velocity profile $u(t)$:

$$u(t) = \begin{cases} u_{on} \sin\left(\pi \dfrac{t}{T_{for}}\right) & \text{for} \quad t < T_{for} \\[2ex] -u_{off} \sin\left[\dfrac{\pi}{T_{back}}(t - T_{for})\right] & \text{for} \quad t \geq T_{for} \end{cases} \qquad (2)$$

2.1.2. Ruessink [RUE]

The RUE parameterization was derived from extensive field measurements of non-breaking and breaking waves for distinct wave climates and beach typologies [18]. The RUE predictor uses the Ursell number (Ur) and empirically derived coefficients to estimate the Skewness (Sk) and Asymmetry (As). The Sk and As are then used to compute the non-linearity r and ϕ terms that are used in the intra-wave $u(t)$ relation of Abreu et al. [47]:

$$u(t') = U_w\, f\, \frac{\sin(\omega t') + \dfrac{r\sin(\phi)}{(1+f)}}{1 - r\cos(\omega t' + \phi)} \tag{3}$$

where $f = \sqrt{(1-r^2)}$ is a dimensionless factor to match the amplitude of u and U_w, and ω is the angular frequency. In addition, t is modified into t' to ensure $u(0) = 0$ m/s.

2.2. Sediment Transport Prediction

The sediment transport in van Rijn [44,45] (VR04) is divided into four components: (1) current-related bed load ($S_{c,b}$); (2) current-related suspended load ($S_{c,s}$); (3) wave-related bed load ($S_{w,b}$); (4) wave-related suspended load ($S_{w,s}$). While we will focus mainly on the wave-related components ($S_{w,b}$) and ($S_{w,s}$), the orbital velocities do affect the current-related transports through the combined shear stress of currents and waves [43,48]. Rather than providing a detailed review of the VR04 formulation, an overview is given here to guide our interpretations.

The general bed load predictor has similar formulation for waves $S_{b,w}$ and currents $S_{b,c}$:

$$S_b = 0.5\,\rho_s\, d_{50}\, D_*^{-0.3} \left(\frac{\tau'}{\rho}\right)^{0.5} \frac{max(0, \tau' - \tau_{cr})}{\tau_{cr}}$$

being

$$\tau' = \frac{1}{2}\,\rho\, f'\, u^2 \tag{4}$$

where S_b = instantaneous bed load transport [kg/m/s]; ρ_s = sediment density [kg/m^3]; ρ = fluid density [kg/m^3]; f' = combined wave and current-related friction coefficient [-]; D_* = dimensionless particle size incorporating sediment and fluid density and viscosity [-]; d_{50} = median sediment grain size [m]; u = instantaneous velocity due to current and waves at reference height [m/s]; τ' = instantaneous grain-related bed-shear stress due to currents and waves [N/m^2]; τ_{cr} = critical bed-shear stress based on the Shields criterion [N/m^2]; For the intra-wave bed load component, S_b uses $u(t)$ derived from IH or RUE to compute the bed-shear stress within the wave period which is then integrated over the wave period to compose the wave-related bed load transport, i.e., $S_{b,w} = \int_{t=0}^{T} S_b\, dt$. The inclusion of a critical shear stress for the initiation of motion means that mobility varies with the degree of non-linearity. The VR04 predicts bed load transport in relation with flow velocity to a power of roughly 3 in high mobility, but much higher powers when the instantaneous shear stress (τ') barely exceeds the critical shear stress (τ_{cr}).

The suspended load transport due to waves included in Delft3D [43,45,49] is calculated based on the parameterized maximum onshore (u_{on}) and offshore (u_{off}) orbital velocities rather than an intra-wave calculation. Furthermore, wave-induced streaming velocities are included and multiplied by the suspended sediment concentration (c) and a phase lag constant f_p, here 0.1

$$S_{s,w} = f_p \left(\frac{u_{on}^4 - u_{off}^4}{u_{on}^3 - u_{off}^3} + u_s \right) \int_a^{3\delta_s} c\, dz \tag{5}$$

where $S_{s,w}$ = suspended load due to waves [kg/m/s]; u_s = wave-induced streaming velocity near the bed; a = reference height [m]; δ_s = wave boundary layer thickness [m].

The current-related suspended load $S_{s,c}$ is computed by the advection-diffusion equation, using bed-shear stress and eddy viscosity to calculate reference sediment concentration by a relation coupled to the bed load, the concentration profile above the bed and flow velocities as the advection term [43]. For calibration purposes, each transport component is multiplied by a user-specific value, with one being the default value.

2.3. Numerical Modelling

Numerical modelling in Delft3D was applied to two well-studied complementary coasts: (1) Katwijk, The Netherlands and (2) Duck, NC, USA. Those locations comprise long-term data of beach profiles and wave measurements in two different wave climates. Duck on the East Coast of the USA represents oceanic wave conditions with swell and locally generated wind-waves, while Katwijk on the Central Dutch Coast, facing the semi-enclosed North Sea, has locally generated sea waves only. Rather than a direct and strict comparison with measured data, these contrasting cases are used to study model behaviours and long-term results. The most important indicator of model behaviour will be the long-term morphological development as a result of net sediment transport trends in comparison with the measured envelope of morphology over the past decades.

The simulations target the nearshore coastal evolution dominated by wave action on a timescale of decades. Over this time, the morphology will be considered in quasi-equilibrium conditions with variations within a bed level envelope. The model will be evaluated on the tendency to develop towards a cross-shore equilibrium condition rather than unrealistic decadal rates of coastal accretion or erosion. In order to make the computational effort feasible, the measured wave climate was reduced to synthetic boundary conditions. In addition, the measured beach profiles were averaged over time and space for the initial bathymetric condition. Tides were included to provide water level oscillations that affect wave propagation and nearshore wave-induced currents. Below we describe the modelling scenarios, the initial beach profiles, wave climate and general model settings.

2.3.1. Modelling Scenarios

To unravel the differences and effects of orbital velocities parameterizations on hydrodynamics, sediment transport and long-term morphology we performed 20 model combinations between Katwijk and Duck (Table 1; 10 scenarios of IH and RUE). We first performed hydrodynamic simulations without morphological updates (constant bed level profile) for the average wave condition of each site (scenarios 1 and 7; see wave condition 7 in Table 2). Here we compared the intra-wave orbital velocities (wave shape), peak velocities u_{on} and u_{off} magnitudes and their difference (Δ *vel*), skewness (Ru) and asymmetry (Ra) coefficients at Katwijk and Duck. We then computed the annual equivalent sediment transport for cross-shore and alongshore directions incorporating the wave climate on a fixed bed for Katwijk and Duck (scenarios 2 and 8). Based on the annual cross-shore-integrated sediment transport we defined calibration values, individually for Katwijk and Duck, aiming: (1) shoreline equilibrium by means of equal cross-shore-integrated onshore and offshore transport magnitudes and (2) same net cross-shore transport magnitude for IH and RUE in addition of equal net bed load and suspended load within the methods. With default annual cross-shore sediment transport we can assess the sediment budget (balance) differences between IH and RUE and how far they are from predicting shoreline stability, when the onshore and offshore magnitudes should be equal to zero. In addition, we want to define calibration factors for sediment transport when IH and RUE would predict the same amount of onshore and offshore transport (and therefore shoreline stability) as well as bed load and suspended load, as they behave differently. In this way we can compare local differences in morphological evolution for the same amount of sediment transport, when only local sediment gradients differ between methods. Full morphodynamic simulations for the wave climate were performed with default and calibrated sediment transport values for Katwijk and Duck, both on an alongshore uniform bathymetry based on the measurements. In addition, we run a scenario with inverted calibration at Katwijk, when the IH parameterization runs with RUE sediment transport calibration factors and vice-versa. Finally, to assess the effects on the alongshore sediment transport, a scenario with calibrated sediment transport values was run with a coastal hump added on the Katwijk coast. For the hump scenario we choose to run only the calibrated scenarios in order to isolate cross-shore processes, as much as possible, when the hump diffuses based on alongshore sediment transport gradients. Here, we assess the volumetric and cross-shore profile evolution of the hump and adjacent areas.

Table 1. Overall summary of performed numerical simulations. Wave conditions are described in Table 2 being wave 7 the single-averaged wave condition and wave climate refers to all schematized waves (Table 2); Default and calibration stand for sediment transport multiplication factors, being default equal to one and calibration values are presented in Table 3. All scenarios were performed for IH and RUE parameterizations.

	Scenarios	Wave Boundary	Bathymetry	Sed. Transport	Morphology
1	Katwijk 1	single wave	uniform	off	off
2	Katwijk 2	wave climate	uniform	default	off
3	Katwijk 3	wave climate	uniform	default	on
4	Katwijk 4	wave climate	uniform	calibrated	on
5	Katwijk 5	wave climate	uniform	inv. calibrated	on
6	Katwijk 6	wave climate	hump	calibrated	on
7	Duck 1	single wave	uniform	off	off
8	Duck 2	wave climate	uniform	default	off
9	Duck 3	wave climate	uniform	default	on
10	Duck 4	wave climate	uniform	calibrated	on

2.3.2. Initial Beach Profiles

The initial bathymetry for the Dutch coast is a time-spatial averaged nearshore bed profile (Figure 1A) obtained from the Dutch JARKUS database [50] near Katwijk. The profiles were measured between 1965 and 2010 along 31 alongshore-distributed profiles covering 15 km of coastline from the dunes up to approximately 18 m depth. The average profile has a slope of 1V:185H up to 2.3 m depth, steepening to 1V:63H up to the mean water line. The sediment grain size composition varies between 200 and 350 μm [5]. A comprehensive morpho-sedimentary description of the Dutch coast is presented in Wijnberg and Terwindt [51] and van Rijn [5]. Similarly, we averaged profiles from 1997 until 2008 (Figure 1B) at Duck situated north of the Field Research Facility Pier. The beach profile at Duck is steeper than at Katwijk. The average profile has a slope of 1V:120H up to 2.4 m depth, steepening to 1V:25H up to the mean water line. The sediment grain size varies between 125 and 250 μm along most of the submerged profile with a second grain size mode of gravel-sized material near the shoreline [52]. An extensive site and data description for Duck is provided in Trowbridge and Young [53] and Gallagher et al. [54]. As can be seen in Figure 1B, the profile was linearly extended from 15 m depth up to the 17 m contour from where the wave data was retrieved. The cross-shore profile covers approximately 5 km at Katwijk and 3.8 km at Duck. From the measured envelopes in Figure 1, most morphological activity is constrained between the dunes and the 6 m depth contour at both sites.

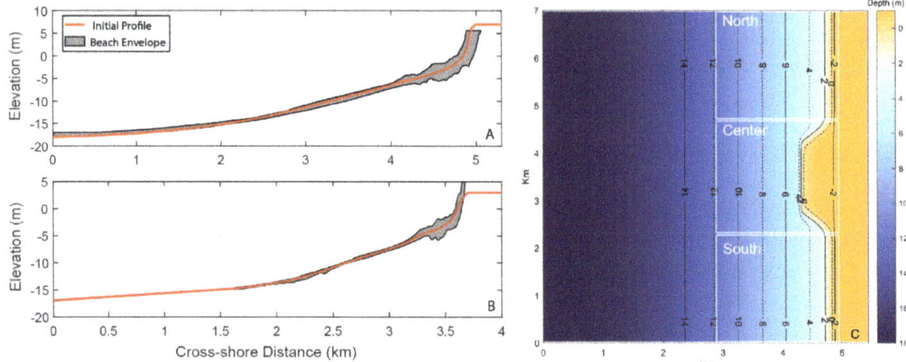

Figure 1. Initial time-space averaged modelling bathymetries and measured beach profiles represented as the overall beach envelope. Measured minimum and maximum coastal profile depths (gray) at Katwijk, the Netherlands (**A**), and Duck, USA (**B**). The smoothed average bed levels (red) correspond with the maps of uniform bathymetry in alongshore direction and were implemented into the numerical model. (**C**) Implementation of an idealized coastal hump into the Katwijk Dutch coastal profile.

The averaged profiles were extended seven kilometers in the alongshore direction creating an initial alongshore uniform bathymetry that suppresses alongshore gradients for the area 2DH models. To study effects of the wave parameterization on alongshore direction, a 3.7 Mm^3 coastal hump was included in further simulations in the Katwijk model setup (Figure 1C; scenario 6). The hump resembles a beach nourishment extending from the dry beach up to approximately 5 m depth. For analysis, the nearshore area is divided into three sections: The central area corresponds to the hump and the other two are the adjacent coasts to the North and South. The computed volumes considered all areas extending from the dry beach into approximately 12 meter depth contour, enclosing the active beach profile, and therefore the majority of the sediment exchange happens via alongshore transport.

2.3.3. Wave Climate

The wave conditions for the Dutch Coast were recorded by the IJmuiden 'Munitiestortplaats' directional buoy in the North Sea between 1990 and 2016. The wave measurements from Duck were recorded by a wave rider buoy (station wvrdr630) from 1997 to 2018. The wave directions were rotated 30° at Katwijk and 161° at Duck, respectively, to realign the waves with our idealized coast based on the local shoreline orientation (Figure 2A,C). Therefore wave directions will be given in this new frame of reference, with 270° implying shore-normal wave incidence.

The wave climate at Katwijk consist of short period waves averaging 4.6 s with 11.7 s maximum. The maximum wave height reaches 7.6 m while the average is 1.32 m. The wave direction has two main components from SW (200°) and NW (310°) but also has a significant frequency of parallel and offshore going waves for 13% of the year. Duck has longer period waves up to 25 s and an average of 8.7 s. The maximum wave height reaches 8.12 m, averaging 0.99 m. The direction has one mode from 300° and limited offshore and parallel going waves summing 1% of the year.

For optimizing computational effort, the original time series were reduced into 13 representative waves following Walstra et al. [55] and Benedet et al. [56]. The wave reduction consisted of four directions and three heights plus an average wave condition to replace the duration of offshore and parallel directed waves. The representative wave conditions were obtained by dividing the wave climate into 12 bins of equal energy E, being $E \sim H_s^2$. Then, the average wave period, direction and the total duration were calculated within each bin. Consequently, one wave condition energetically represents a range (Figure 2B,D), consisting of significant wave height, wave period, wave direction and duration. The reduced wave climate for Katwijk and Duck is shown in Table 2, with condition number 7 being the averaged condition to replace the recorded offshore and parallel going waves.

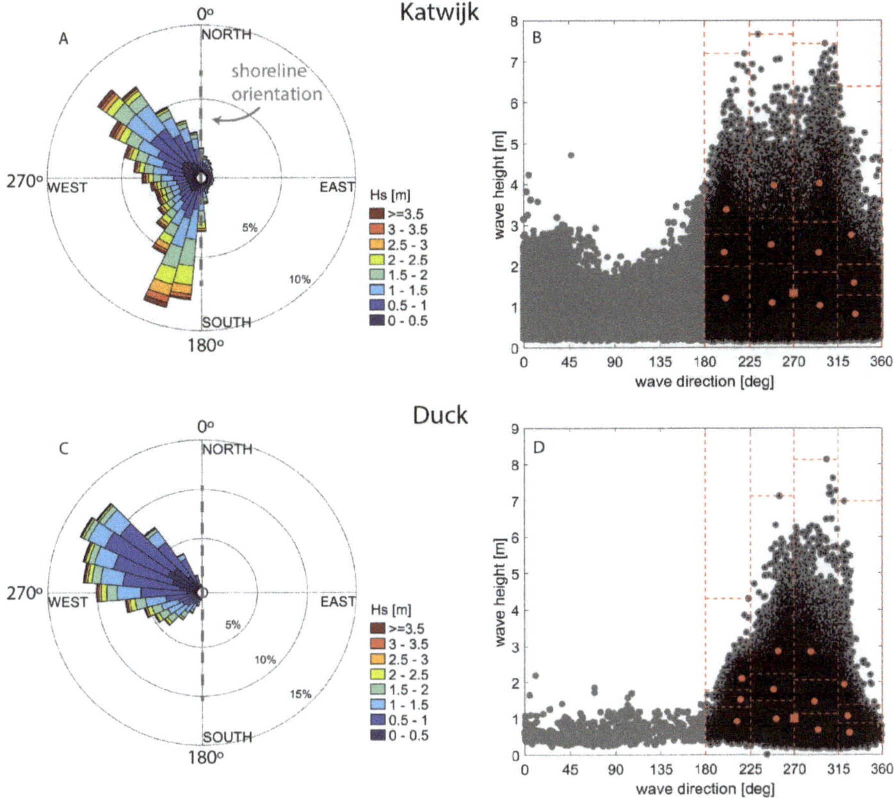

Figure 2. Measured and reduced wave climate at Katwijk and Duck. (**A,C**) Measured wave polar histogram for significant wave height (H_s); (**B,D**) and the reduction to wave bins for modelling. Data for Katwijk (**A,B**) recorded at the IJmuiden 'Munitiestortplaats' between 1990 and 2016. Directions were rotated 30° for application in the model domain. (**C,D**) Data for Duck collected near the FRF Pier from 1997 to 2018 after 161° rotation. (**B,D**) Measured wave heights (gray) were filtered to selected onshore-directed cases (black). Red lines divide each wave bin into equal energy and red dots are reduced wave conditions that represent the quadrant. The red square represents the average condition in the shore-normal propagation direction. Reduced wave conditions are presented in Table 2.

The 13 reduced wave conditions were organized as a constant alternating time-series sequence, following the order from Table 2, changing every hydrodynamic hour until its duration was exhausted. In this approach we are constantly changing the wave energy followed by wave direction. While the ordering of wave conditions is known to affect the dynamics of surfzone bars [13,55], the present focus is on equilibrium morphology and large scale sediment transport behaviour. This alternation process ensures that no wave condition perpetuates for too long driving the morphology into an alternative equilibrium.

Table 2. Reduced wave climate for Katwijk and Duck to 13 conditions of wave height (H_s), period (T_p), direction (*Dir*, rotated to the model domain) and normalized duration (%). Wave 7 is the average condition that replaces the offshore-directed and shore-parallel wave conditions.

	Katwijk—NL					Duck—USA			
Wave	Hs	Tp	Dir	Duration	Wave	Hs	Tp	Dir	Duration
	m	s	deg	%		m	s	deg	%
1	1.20	4.1	201	20	1	0.92	4.6	211	4
2	2.34	5.2	200	5	2	1.53	5.5	215	1
3	3.37	6.0	202	3	3	2.10	6.2	216	1
4	1.08	4.2	248	11	4	0.98	7.7	252	18
5	2.53	5.5	247	2	5	1.80	7.7	249	5
6	3.96	6.5	250	1	6	2.87	9.2	254	2
7	1.32	4.6	270	13	7	0.99	8.7	270	1
8	1.02	4.7	297	20	8	0.68	9.8	294	43
9	2.33	5.7	296	4	9	1.47	9.8	291	9
10	4.02	6.9	296	1	10	2.85	11.2	287	2
11	0.81	4.2	333	14	11	0.61	7.2	326	9
12	1.58	4.9	332	4	12	1.06	7.5	325	3
13	2.76	5.9	329	1	13	1.94	9.5	321	1

2.3.4. General Model Configurations

The numerical simulations are depth-averaged 2DH models with cross-shore resolution varying from 50 m offshore to 12.5 m towards the coastline and 50 m on the alongshore direction. The hydrodynamics are coupled with wave conditions computed with SWAN as online communication of bathymetry, waves, currents and water level. In combination with waves, the hydrodynamics for sediment transport are solved in Generalized Lagrangian Mean (GLM) mode to account for wave-induced processes, e.g., Stoke's drift [43].

The propagated wave conditions were coupled with flow every 30 min in order to consider the wave–current interaction, wave-induced currents, water level fluctuations due to tides and morphological bed evolution. Based on available tidal records, the tidal water level amplitude was set to 1.0 m at Katwijk and 0.5 m for Duck, both as 12 h harmonic tides. To avoid undesired boundary effects, the alongshore domain encloses a 7 km long coastline with null gradient Neumann boundary conditions at the cross-shore boundaries.

After sensitivity analysis, not shown here, a constant morphological acceleration factor (morfac) between 10 and 20 was used to speed up the simulations performing in total 10 morphological years. The simulations 3 and 9 were performed with morfac equal to 10; the hump scenarios with morfac equal to 12; and models 4–6 and 10 with morfac equal to 20. These morfac values are similar to those used in other studies (e.g., [22,27,37]). Sediment transport was limited to depths great than 0.3 m at Katwijk and 0.4 m at Duck based on the non-dimensional wave period condition $T_p \sqrt{g/h} < 40$, following Ruessink et al. [15], to exclude swash zone processes. A single sand fraction of 250 μm was applied in all model runs. Other model settings are specified in the Supplementary material.

3. Results

3.1. Cross-Shore Orbital Velocities

The cross-shore distribution of orbital velocities predicted with RUE and IH over the Katwijk and Duck profiles are shown in Figure 3A,B,F,G for the average wave condition (scenarios 1 and 7 in Table 1; wave 7 in Table 2). IH does not explicitly computes Sk and As, then for direct comparison between methods we computed skewness (R_u) and asymmetry (R_a) coefficients based on velocity and acceleration, respectively (see Equation (1)). Here IH consistently predicts higher skewness than RUE, and also predicts skewness in deeper water (Figure 3D,I). For example, the skewness at Katwijk starts around 14 m depth for IH and 8 m for RUE. This effect is more pronounced at Duck where the wave period is larger. At Duck, IH is already skewed from the seaward boundary, at 17 m depth contour, while RUE shows linear behaviour up to 14.5 m depth. RUE predicts lower values of skewness seaward of the surfzone and develops asymmetry towards the surfzone, corresponding to 3 m depth at Katwijk and 6 m at Duck. The interval asymmetry exceeds skewness correspond with the zone of wave breaking at 1.5 m and 2.5 m depth at Katwijk and Duck, respectively (Figure 3E,J). On the other hand, IH has skewness-only, which increases strongly towards the waterline.

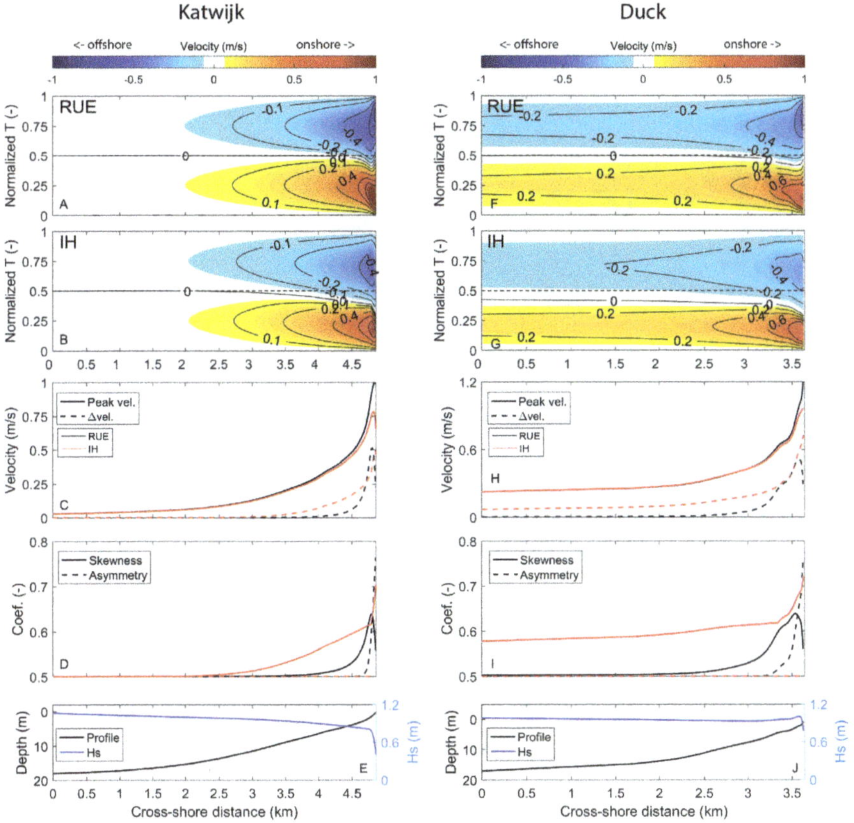

Figure 3. Hydrodynamic comparison between RUE and IH behaviour for the average wave condition (Wave 7) along the beach profiles of Katwijk ((**A–E**) left) and Duck ((**F–J**) right). Orbital velocities (positive onshore and negative offshore) within the wave period T for RUE (**A,F**) and IH (**B,G**). (**C,H**) Peak orbital velocity and the difference between onshore and offshore velocity amplitude $\Delta vel = |u_{on}| - |u_{off}|$, color-coded for RUE and IH. (**D,I**) Skewness and Asymmetry coefficients. (**E,J**) Beach profile and significant wave height.

The main effects of the skewness predicted with IH into the orbital velocities and wave shape are twofold: (1) lower offshore-directed velocities in comparison with the onshore velocities and (2) the shorter crest period marked by the zero contour line deviating from half of normalized wave period. For RUE the sinusoidal linear shape persists longer towards the shoreline so that the onshore and offshore peak velocities are of similar magnitudes, deviating when skewness develops during shoaling process and again become nearly equal when asymmetry develops in the surfzone (Figure 3C,H). The peak onshore velocity is similar for both methods except in the very shallow areas where RUE predicts higher magnitudes. This means that the onshore velocities are similar for IH and RUE and the lower prediction of offshore velocities by IH is the main difference concerning the peak velocities.

These effects result from how IH translates the non-linearity into the orbital wave shape. After computing the peak onshore and offshore velocities the wave period is split artificially based on the ratio between (u_{on}) and (u_{off}). The intra-wave velocity is then computed separately for each direction with a sine expression (Equation (2)). This leads to a discontinuous wave shape for the intra-wave orbital velocity (Figure 4). As a consequence, IH produces larger skewness than RUE, especially for longer period waves as demonstrated in Figure 3 at Duck.

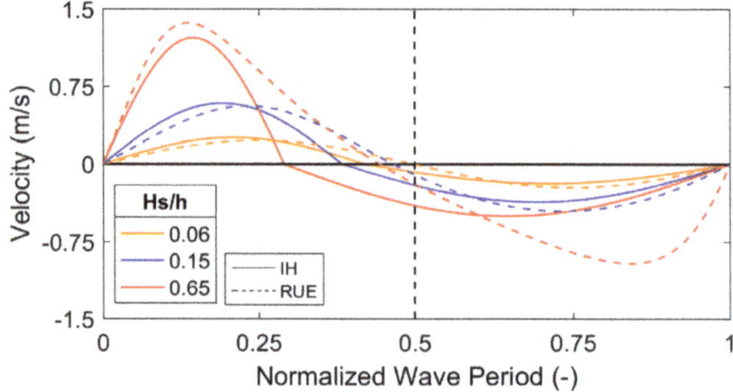

Figure 4. Intra-wave velocity profile (wave shape) computed at Duck with IH and RUE along different profile locations from deep (orange) to shallow (red) water represented by the ratio of wave height divided by water depth H_s/h. Positive velocities are onshore-directed.

3.2. Sediment Transport on Alongshore Uniform Coasts

The orbital velocities are coupled to VR04 sediment transport via combined current and wave bed-shear stress. For the wave-related bed (Equation (4)) and suspended load (Equation (5)), the difference between onshore and offshore velocities is the main driver of cross-shore sediment transport and therefore we consider VR04 a skewness-based sediment transport predictor. For the bed load component, the critical shear stress enhances the non-linear behaviour in sediment transport near the beginning of motion due to the division of excess bed-shear stress by the critical shear stress.

Thus, the relation between flow velocity and sediment transport, in addition of critical shear stress of motion, creates a rather complex relation when adding non-linear oscillatory wave processes. To unravel this phenomena, Figure 5 compares the results of sediment transport (combined cross-shore and alongshore) between IH and RUE for each wave condition from Table 2 integrated over a tidal cycle. All IH simulations predicted more sediment transport than RUE. The sediment transport factor, defined as IH/RUE, varies from 1 to 3.1, except for the wave-related bed load transport ranging from approximately 3 to 12 with an outlier of 29. In general, the largest deviations between methods derive from small sediment transport values associated with low-energy frequent-duration wave conditions, specially for the wave-related bed load (Figure 5A). To reinforce this trend, the observed outlier corresponds to the average wave condition at Katwijk ($Hs = 1.32$ m; $Tp = 4.6$ s) due to a relatively large difference in transport magnitude while the absolute values were small (1.08×10^{-4} m^3 for IH and 3.72×10^{-6} m^3 for RUE). Therefore, the largest discrepancies between methods derived from situations near the beginning of motion instead of from more energetic (stormy) wave conditions. As these low energetic conditions are predominant in the yearly wave climate, they may result in large deviations for the annual sediment budget.

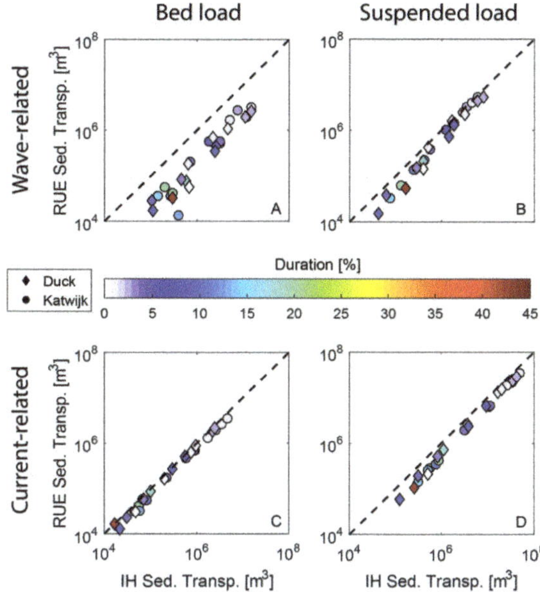

Figure 5. Sediment transport computed with default settings for individual wave conditions integrated over the cross-shore profile and tidal cycle for IH and RUE with colors representing the duration of each wave condition in the reduced wave climate. (**A**) wave-related bed load; (**B**) wave-related suspended load; (**C**) current-related bed load; (**D**) current-related suspended load. The dashed line represents a 1:1 reference.

3.2.1. Cross-Shore Sediment Transport

Next we computed the annual equivalent cross-shore sediment transport (scenarios 2 and 8 in Table 1) on the alongshore uniform coast applying the reduced wave climate at Katwijk and Duck. First we computed the annual bed load, suspended load and total transport of default sediment transport factors (Figure 6A,B,E,F) and afterwards we defined calibration factors (Figure 6C,D,G,H and Table 3) aiming stable nearshore cross-shore profile for the morphological simulations.

At both sites, with default sediment transport settings, the net cross-shore annual bed load (combined of current and wave) is onshore-directed while the suspended load is offshore-directed. As individual components, not shown here, the wave-related bed load ($S_{w,b}$) and suspended load ($S_{w,s}$) transports are onshore-directed while the current-related bed load ($S_{c,b}$) and suspended load ($S_{c,s}$) are offshore-directed. The IH simulations shows higher magnitudes for all sediment transport components. The higher onshore-directed bed load leads to an onshore-offshore sediment budget imbalance, which in this case will result in large shoreline accretion, mainly caused by the wave-related bed load component (Figure 6B,F). The annual cross-shore-integrated wave-related bed load transport is 4.62 times higher in IH than in RUE for Katwijk and 6.37 times higher for Duck. The wave-related suspended load is a factor of 1.29 and 1.64 higher for Katwijk and Duck, respectively. The choice of orbital velocity parameterization also affects the current-related transports through the computation of sediment concentration from the combined wave–current shear stress; the current-related bed load was 1.41 times higher and the suspended load 1.37 times higher. After combining the four transport components, the total net cross-shore transport is onshore-directed by a factor of 4.8 higher in IH than RUE at Katwijk and 7.8 at Duck.

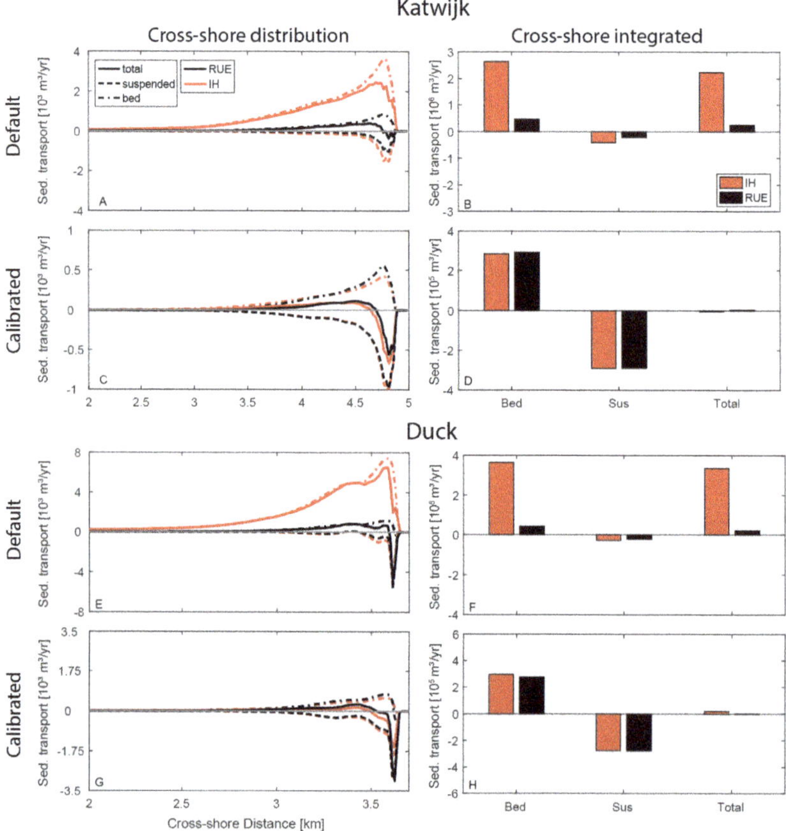

Figure 6. Annual cross-shore sediment transport at Katwijk (**A–D**) and Duck (**E–H**). Positive values are onshore-directed. IH and RUE are color-coded with red and black. Note the difference in vertical axis scale between plots. (**A,C,E,G**) shows the cross-shore profile of sediment transport and (**B,D,F,H**) integrates the values in the cross-shore direction. Panels (**A,B,E,F**) shows sediment transport values for default sediment transport factors and panels (**C,D,G,H**) after calibration is applied (see Table 3).

Table 3. Sediment transport calibration factors for Katwijk and Duck for IH and RUE formulation. w stands for wave-related; c for current-related; b for bed load; and s for suspended load.

	Calibration Katwijk		Calibration Duck	
	IH	**RUE**	**IH**	**RUE**
$F_{w,b}$	0.155	0.720	0.112	0.720
$F_{w,s}$	0.154	0.200	0.123	0.199
$F_{c,b}$	0.709	1.000	0.875	1.000
$F_{c,s}$	0.272	0.375	0.228	0.360

Based on the annually cross-shore-integrated sediment transport differences between IH and RUE, we performed a sediment transport calibration to be applied on the morphological models in addition of default value scenarios. Recapping, with this calibration procedure we want to achieve nearshore profile stability, with minimum shoreline translation, and equal total net cross-shore sediment transport between IH and RUE. First, we target the net total cross-shore transport to be zero for each method (i.e., IH and RUE). We started keeping the current-related bed load of RUE equal to one, as it computes the least transport magnitudes (Figure 5C), followed by correcting the wave-related bed load, wave-related suspended load and finally the current-related suspended load. This first iteration achieves the onshore-offshore balance of sediment and a second iteration was made to match the annual total transport between IH and RUE. Applying those calibration factors (see Table 3) resulted in equal annual cross-shore-integrated magnitudes of bed load and suspended load with net total load nearly zero (Figure 6D,H). Based on the necessary calibration factor to correct sediment transport, the largest differences are found in the wave-related bed load sediment transport, so that the $F_{w,b}$ was reduced to 11.2% and 15.5% of the original magnitude for IH at Duck and Katwijk, while RUE was kept at 72%. The $F_{w,s}$ was similarly reduced around 20% as the wave-related suspended load does not fully account for the intra-wave sediment transport and actually has similar behaviour in direction and magnitude as the wave-related bed load. Finally the $F_{c,s}$ was reduced between 23% and 37% to close the cross-shore sediment budget. After calibration, the cross-shore distribution of sediment transport shows similar shapes and gradients (Figure 6C,G), except in the shallowest areas where we showed deviations starting from the hydrodynamics (Figure 3). Therefore, local morphological variations are expected along the profile, especially in the surfzone and intertidal areas.

3.2.2. Alongshore Sediment Transport

The alongshore-directed sediment transport rates were analyzed for Katwijk, only, in order to provide insights on the hump evolution simulated with the same settings. Both IH and RUE methods predict net northward sediment transport at Katwijk. For default sediment transport settings IH and RUE predict 1,320,000 m^3/year and 915,000 m^3/year (Figure 7A,B). Thus, IH predicts 405,000 m^3/year (that is 1.44 times) more net northward sediment transport compared to RUE. The bed load sediment transport factors for waves and currents are 2.0–2.1 and the suspended loads 1.0–1.3. However, after applying the cross-shore based calibration (Table 3), IH results in much lower values for the net alongshore transport, namely 73% of reduction from the default value; and 56% of reduction for RUE (Figure 7C,D). In volume, that represents a reduction of 968,200 m^3/year for IH after calibration while for RUE 517,000 m^3/year. This means that the sediment transport calibration aiming coastal stability forces changes in the alongshore sediment transport with more than a factor of two for IH and slightly higher for RUE. This reduction of littoral drift may affect long-term morphological development, and the timescale thereof, along non-uniform coasts.

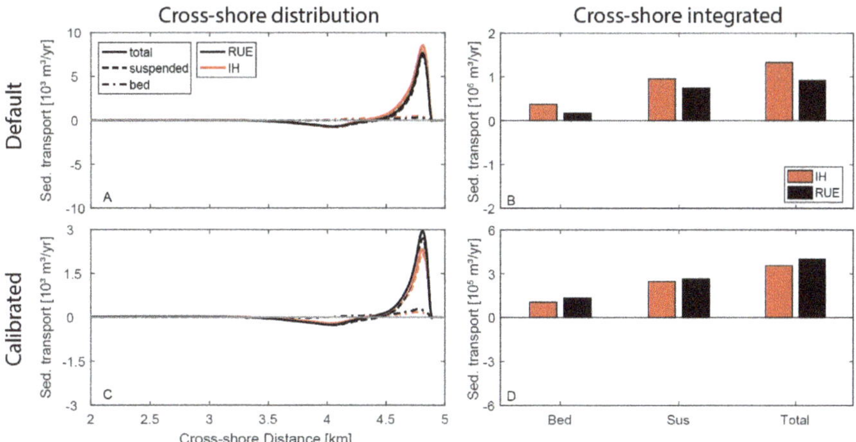

Figure 7. Annual alongshore sediment transport at Katwijk. Positive values are northward. Note different vertical axis scales. (**A,C**) cross-shore profile of net alongshore sediment transport. (**B,D**) Integrated the values in the cross-shore direction. Panels (**A,B**) show default sediment transport values and panels (**B,D**) show calibrated IH and RUE (see Table 3).

3.3. Alongshore Uniform Coastal Morphology

Ten morphological years of simulation were performed with default and calibrated transport factors (Figure 8; scenarios 3, 4, 5, 9, 10 in Table 1). For the default sediment transport simulations at Katwijk, the wave-related transport appears from approximately 12 m depth upwards. Between 12 m and 5 m IH erodes the bed creating net onshore sediment transport. As a result, the shallower areas and the shoreline became excessively obese and steep. Simulations with RUE also show net onshore transport, though less than IH, but hardly eroding deeper areas. In comparison to the measured bed profile envelope, IH prediction is outside the envelope from 9 m depth up to the shoreline while RUE deviates only around the intertidal zone. From the measured profile envelopes we expected most morphological dynamics happening from 6 m up to the shoreline, therefore, especially IH overestimated the profile changes.

Following calibration, both IH and RUE stayed within the envelope boundaries, with IH showing slightly more deposition in the intertidal shallow area. In addition, the calibration succeeded in balancing the onshore and offshore transport magnitudes towards a stable profile with minimum shoreline translation. The scenarios with RUE factors applied in IH and vice-versa (inverted calibration, scenario 6) show the largest morphological deviations. The RUE simulation with IH factors has the closest fit with the initial (equilibrium) profile which, as a first glance, suggests the best performance. However it is important to highlight that the extreme lower transport factors applied (see Table 3) for wave-related transport (around 10–15%) just shows lower morphological dynamics rather than being the best calibration. On the other hand, IH with RUE predicts the largest shoreline displacement (i.e., 290 m) and large deposition on shallow waters up to 4 m depth. This extreme case is the combination of high onshore transport rates due to wave-related bed load in addition of the reduced offshore transport caused by the smaller factor on the current-related suspended load. Similar response was observed for Duck (Figure 8B) with a total shoreline displacement of 230 m with IH in comparison with 67 m with RUE for default scenarios. After calibration there is a divergence in the intertidal and subtidal areas up to 2 m depth where IH results in a larger deposit. At Duck both methods resulted in more erosion on the lower shoaling zone and deposition on the upper shoaling zone and surfzone.

Despite the parameterization choice and calibration factor, there is a general and alike steepening of the shoreface in all simulations, excepted when the wave-related transport was drastically reduced in the scenario of RUE with IH calibration factors.

The combined interpretation of default, calibrated and inverted calibration unravels: (1) the wave-orbital motion and its parameterization highly affects the sediment transport in the nearshore and consequently the morphological development in this area; and (2) how much the (wrong) calibration factors can affect the morphological development, especially for IH that predicts larger onshore transport and therefore is more sensitive to small changes in calibration factors. The wave-related sediment transport is the main driver of nearshore sediment transport, which promotes only onshore sediment transport for the 2DH configuration. This trend ultimately results in disproportional shoreline progradation, when the offshore-directed current-related transport cannot balance this onshore component, in addition of causing shoreface steepening. In this perspective, IH with larger (in our case 4.6–7.8 times higher) onshore transport is more sensitive and prone to calibration, implying that small inaccuracies in calibration factors result in large profile and shoreline translation.

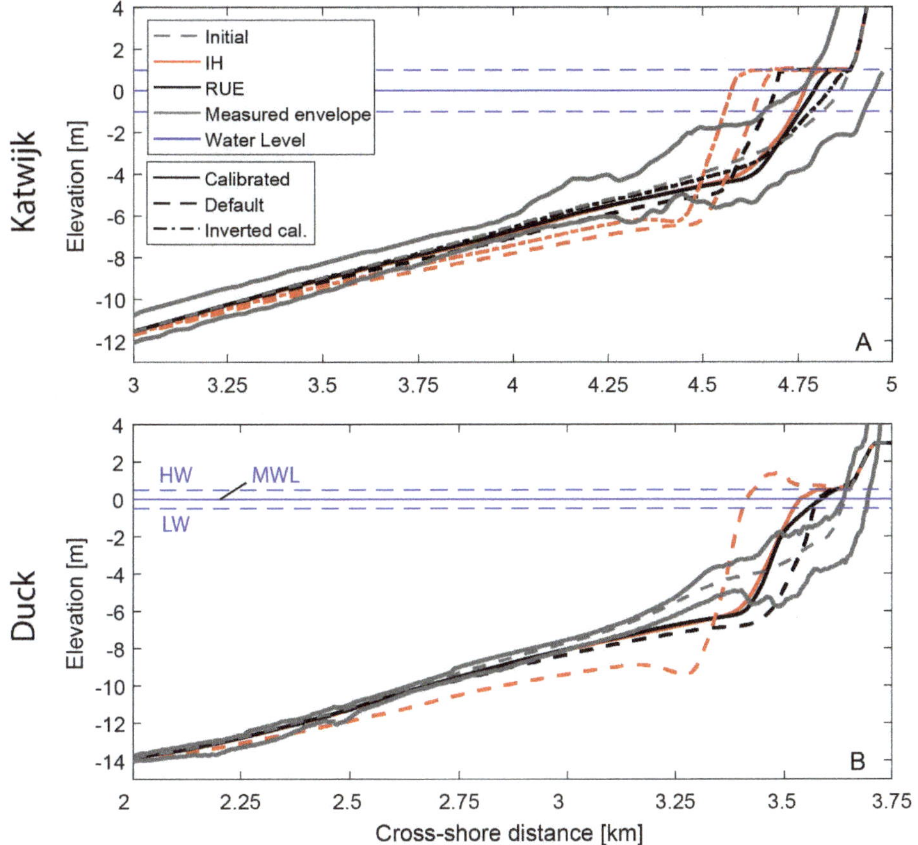

Figure 8. Cross-shore bed profiles at (**A**) Katwijk and (**B**) Duck after 10 morphological years. Katwijk (**A**) contains default, calibrated and inverted-calibration scenarios while Duck (**B**) default and calibrated. Initial and measured envelope are shown in gray, and water levels in blue. IH and RUE are color-coded with red and black.

3.4. Alongshore Non-Uniform Coast Morphology—Coastal Hump

Finally, we performed morphological simulations with a 'hump' (scenario 6 in Table 1) using the calibrated factors (Table 3) in order to exclude, as much as possible, cross-shore effects (i.e., erosion and deposition leading to shoreline translation) that were already extensively described in the alongshore uniform coast setup. As can be seen in Figure 9, erosion and shoreline retreat is concentrated within the hump central area. Most hump sediment is transported and deposited in the North adjacent area while the South benefits less due to the northward net sediment trend (Figure 7).

The volumetric evolution follows pulses of fast diffusion when the more energetic waves occur in the time series followed by calmer period with slower decay (Figure 9). In the Center, RUE predicts faster loss of sediment in comparison with IH. After 10 years, the hump with RUE decayed to 57% from its initial volume of 3.7 Mm^3 while IH decayed to 66%, representing a deviation of 336,400 m^3 between the two methods.

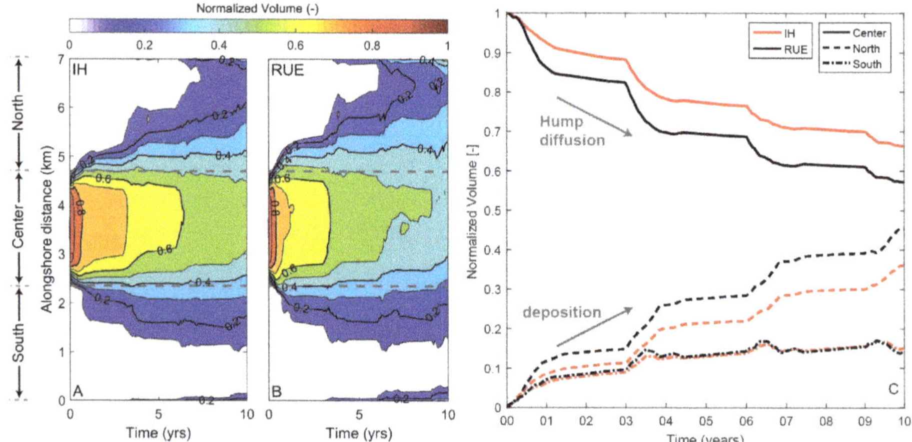

Figure 9. Volumetric diffusion of the coastal hump along the coast within 10 years. (**A**,**B**) Alongshore evolution of the normalized hump volume through time. (**C**) Normalized volume per zone, where Center is on the hump (indicated by dashed lines in a,b and also in Figure 1C) and North and South are the adjacent areas from the hump. The temporal variability is due to the order of wave conditions.

The profile evolution after 10 years shows the retreat of the center profile, faster for RUE, while the North concomitantly responds with larger shoreline progradation when compared to the South (Figure 10). In these scenarios, RUE predicts faster diffusion of the hump itself. This means, for example, that a beach nourishment would feed the adjacent downdrift areas faster, while lasting shorter, in comparison with IH predictions. Therefore, the parameterization choice affects the timescale of alongshore morphological processes. In general, the profiles evolved towards the measured envelope in a trend to restore the alongshore uniform dynamic equilibrium when the alongshore transport gradients tend to zero. Despite the introduction of this large shoreline perturbation, the final cross-shore profile (Figure 10) did not differ in shape from the alongshore uniform case (Figure 8A), which is a further indication that the final morphology is indeed a robust equilibrium condition for the chosen wave climate, calibration and orbital velocity parameterization.

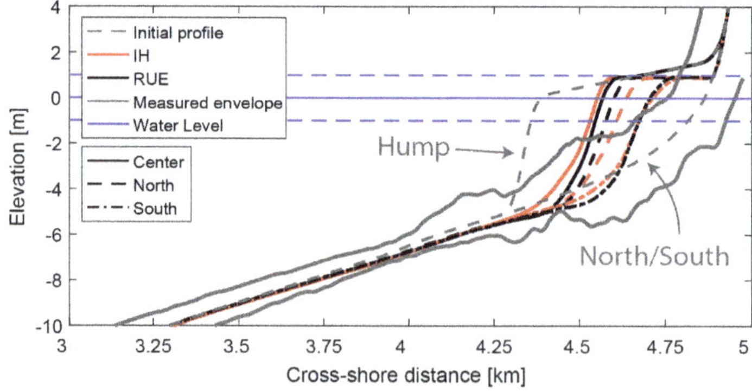

Figure 10. Cross-shore profiles over the hump and adjacent areas at Katwijk after 10 morphological years. Center presents the hump area while North and South are the adjacent areas from the hump (see Figure 1). Measure profile envelope (solid lines) and initial model bed levels (dashed lines) are depicted in gray.

4. Discussion

The near-bed orbital velocities parameterizations differ in predicting wave shape transformation in shallow water and profoundly impact the resulting long-term sediment transport and morphological evolution. Below we discuss the main implication of parameterization choice.

4.1. Long-Term Morphodynamic Evolution

The main consequences of higher onshore-directed sediment transport by IH are twofold: (1) erosion of deeper nearshore areas where waves were not expected and observed to cause such effect (Figures 8 and 11) and (2) overfeed of the shallow upper beach profile. Here, the first effect was caused by non-linearities (skewness) starting earlier (deeper) in the wave propagation path (Figure 3) resulting in a onshore-directed gradient that pushes the sediment towards the shoreline (Figure 6). This phenomena combined with incorrectly increasing skewness in the surfzone (Figure 3) results in larger shoreline progradation (Figure 8) due to excessive deposition, beyond the measured profiles and volumes, and imbalance between onshore and offshore-directed transport (Figures 6 and 11). This excess of onshore sediment transport was larger for the longer period waves of Duck, reaching a factor of 7.8 when considering the annual sediment budget, which translated into unrealistic shoreline progradation and profile evolution (Figure 8B). With default settings, RUE parameterization also overestimated onshore transport, however to less extent in comparison with IH.

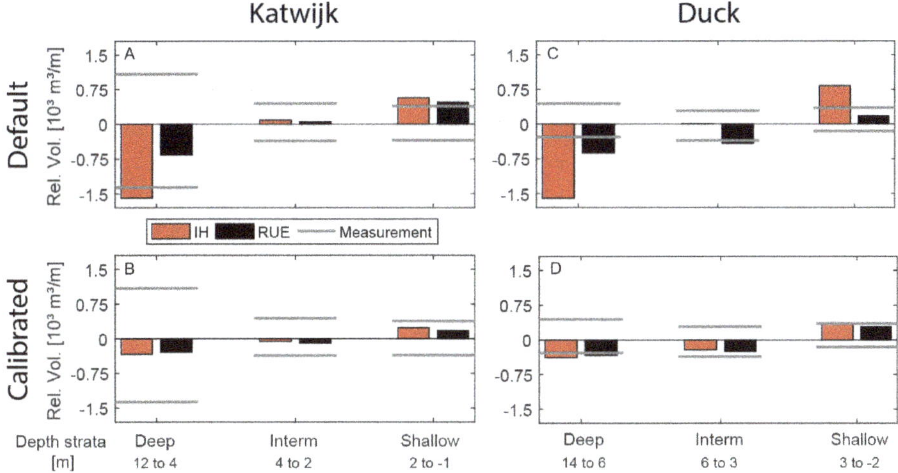

Figure 11. Profile volume after 10 years relative to the initial profile divided in depth zones. Positive values show deposition. Pannels (**A,B**) (left) are results from Katwijk and (**C,D**) (right) from Duck; (**A,C**) (top) are results for default sediment transport and (**B,D**) (bottom) for calibrated scenarios. The gray lines represent the minimum and maximum measured volumes as a reference.

The main reason for unrealistic net onshore sediment transport is the combination of the wave shape parameterization and the importance of the threshold for the beginning of sediment motion in the lowest wave classes. The skewness-only IH method predicts larger skewness and sediment transport in comparison with the skewness-asymmetry RUE predictor. The higher sediment transport factors (i.e., IH/RUE), especially for the wave-related bed load component, correlates with lower steepness and low energy waves (Figure 12), including the outlier (i.e., factor of 29) represented by the average wave condition of Katwijk. In Figure 12 we defined wave power as $Power = \dfrac{\rho g^2 H_s^2 T_p}{32\pi}$ in Watts and wave steepness as $Steepness = H_s/L$. The largest discrepancies between the two methods thus derive from situations near the beginning of motion for the bed load component (Equation (4)) when the sediment transport is strongly non-linear, in this case, higher than a power of 3 for VR04. These low energetic conditions have a larger duration (Figure 5) in comparison with the high-energy stormy events and are the dominant wave conditions in the wave climate (Table 2). Therefore, one can expect sediment transport deviations within a factor of 3–12 when simulating morphology with a wave climate or even higher when including an individual wave condition only, which is commonly of small–intermediate wave energy (e.g., [22,23,30,32–37]).

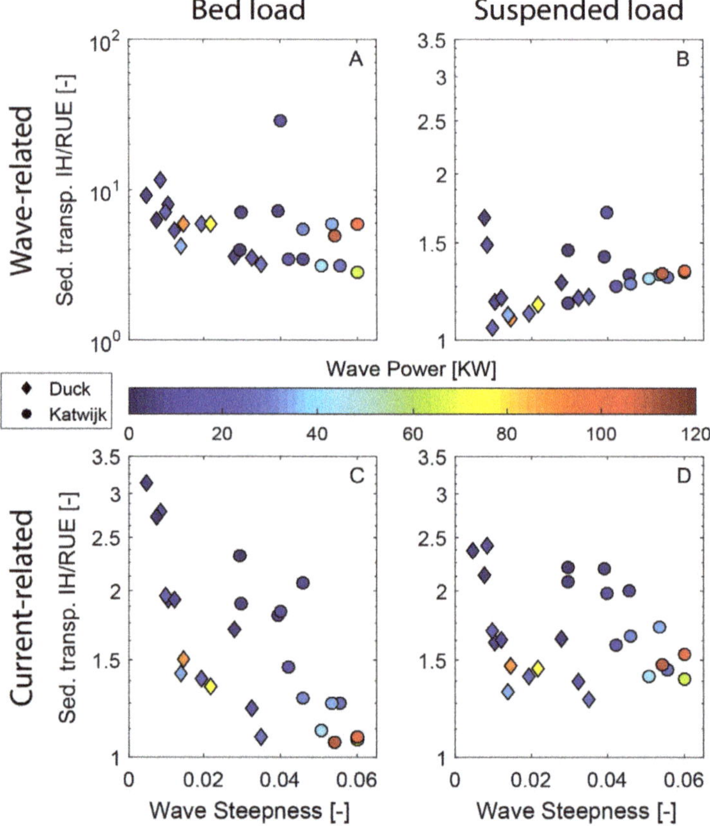

Figure 12. Tidally integrated sediment transport factors of IH/RUE parameterizations against wave steepness at Katwijk and Duck, with colors representing the wave power for each wave case from the climate (Table 2). (**A**) wave-related bed load; (**B**) wave-related suspended load; (**C**) current-related bed load; (**D**) current-related suspended load.

After applying calibration factors on the sediment transport, the overall equilibrium profiles and volume changes with IH and RUE were fairly similar (Figures 8 and 11). Nonetheless there are discrepancies in deeper and especially shallower areas. The overfeeding near the shoreline was also observed for other skewness parameterizations methods in cross-shore models (e.g., [12,15,17]). Meanwhile, the shoaling zone, where skewness dominates, showed closer results. However, to achieve these conditions, the wave-related sediment transport for IH had to be reduced to 10–15% from the default values. Such reduction, although extreme, is not an exception when compared to recent studies [23,27,37], which raises serious doubts about the validity of IH to coastal environment applications in combination with general sediment transport predictors (e.g., [44,45,57]). Based on the assumption that 2DH models have to compensate for the lack of offshore-directed transport, with IH we are implicitly considering that 85–90% of transport is offshore-directed due to return currents, for example. On the other hand, with RUE, when we applied a correction of 70%, we are compensating for 30% of offshore-directed transport. The latter figure better represents an average condition between calm and storm conditions, while the 85–90% of IH represents storm conditions when undertow is dominant over short-wave non-linearity [7,12,54].

The main adverse consequence of (over)compensating the sediment transport aiming to counteract shoreline translation is the drastic reduction of alongshore transport, in the order of 73% in our case with IH. This highlights the current limitations in properly reproducing overall morphology and time-scales in one model. Reducing the net alongshore transport by a factor of 0.73 (i.e., 968,000 m^3/year) represents 385% of the annual net alongshore transport computed by van Rijn [5], based on 200,000 m^3/year estimation for the central Dutch Coast. This means that in applications of IH, one could either realistically simulate the alongshore process or the cross-shore process but not the combination. This limitation further hampers the hindcast and forecast of more complex environments, like tidal basins and inlets which work in strict balance and feedback between waves, tides and fluvial processes in three dimensions.

4.2. Limitations and Perspectives for Wave-Driven Sediment Transport Prediction

Yet our results derive from the Delft3D model, we believe that the RUE parameterization would also improve other morphodynamic numerical models, not restricted to 2DH applications, as it corrects a fundamental mismatch in wave hydrodynamics, which is common to other 1D-2DHV-3D morphodynamic models. Skewness-only parameterizations proved to be able to reproduce beach recovery, after storms, in specific cross-shore 1D/2DV models (see compilation in [12]). However, as demonstrated in Dubarbier et al. [17], the performance near the shoreline is the result of overcompensation of skewness-related transport, which has been used in the past to compensate for the lack of asymmetry effects in sediment transport functions. Here we observed a similar issue in the comparison of hydro-morphological results with IH versus RUE in a model-to-model approach and validation of results with measured beach-envelopes. Therefore, it is relevant to consider the origin and physical basis of each parameterization. RUE derived from nearshore field measurements, including the Central Dutch and Duck coasts [18], while IH and its further adaptations were based on scaled lab experiments [10,16] intended to reproduce wave-skewness. Therefore, we can argue that RUE provides better estimations of near-bed orbital velocities than IH, especially for our case-studies in Katwijk and Duck as shown by our hydrodynamic and morphological results. The important improvement is that these effects need no longer to be compensated by extreme calibration factors on the sediment transport, which affects alongshore transport as well, but addresses the hydrodynamics of orbital velocities. Consequently, the approach in this paper is suitable for large-scale, long-term 2DH modelling but still limited for predicting coastal beach profiles, as all scenarios predict rather steep and convex shoreface profiles (Figure 8), which we attribute to the lack of (quasi)3D-vertical processes combined with the dominant onshore-directed sediment transport driven by waves, similarly to observations from van Rijn et al. [12] and Grunnet et al. [27].

The IH predictor overestimates sediment transport mainly for the wave-related bed load component (Figures 6 and 12). A consequence is that, IH also promotes morphological diffusion due to bed slope gravitational effects. The bed slope effects on sediment transport and morphological development strongly determine the results for situations with currents-only as demonstrated in Baar et al. [58,59]. With the overestimated bed load contribution of waves, we expect stronger diffusion affecting the very shallow areas represented by shoals, bars and channels. The down-slope transport acts as a natural damping mechanism of morphological perturbations, however, its overprediction leads to unnaturally flat morphology. This damping effect was also explored by Dubarbier et al. [17,60]

for nearshore sandbars where increasing the downslope transport component decreased the growth of nearshore sandbars in their model.

The results generalize to many models and general sediment transport predictors that do not account for asymmetry-driven sediment transport (e.g., [43,45,61]). On the other hand, specialist, coastal-oriented sediment transport predictors such as Dubarbier et al. [17] and van Thiel de Vries [62] incorporated asymmetry-driven sediment transport, but these are strongly dependent on user-defined calibration due to the lack of a robust physical relation [63]. Moreover, such coastal-oriented formulations are poorly tested and arguably unsuitable for fluvial processes that are important in the fluvial-tidal transition that the general transport predictors can cover. Thus, further research should focus on robust incorporation of asymmetry-driven sediment transport into general and broadly applicable morphodynamic models, while the present advance opens up the possibility of long-term morphological modelling of coastal systems where cross-shore and alongshore transport are of similar importance.

5. Conclusions

The parameterization of wave-induced near-bed orbital velocities highly affects the long-term (year to decades) prediction of the nearshore morphology due to its non-linear relations with sediment transport. The comparison of the Isobe Horikawa (IH) [10] skewness-only parameterization versus the skewness and asymmetry method of Ruessink (RUE) [18] within the Van Rijn (VR04) transport equations [43] shows that a better representation of wave shape and near-bed orbital velocities leads to overall more realistic morphodynamic predictions.

The IH parameterization predicts larger skewness and onshore-directed sediment transport in comparison with the RUE method. Depending on the wave condition, the tidally integrated net transport with IH was between 3 and 12 times larger than with RUE, with an outlier of 29. The largest differences were observed for calm wave conditions, which are dominant in the wave climate, further enhancing the difference between the two methods in the yearly sediment budget. Thus, with default sediment transport settings, IH simulations led to an overfeeding of shallow areas while eroding the deeper portion of the profile. RUE simulations with default settings also overpredict onshore-directed transport in the shoaling and surfzone, however, to a lesser extent. After applying calibration factors on the sediment transport components, to ensure shoreline stability by means of equal yearly net onshore and offshore sediment transport, both IH and RUE predicted profiles within measured beach envelopes at two sites selected, i.e., Duck, NC, USA and Katwijk, NL. However, IH wave-related sediment transport needed to be reduced to 10–15%, which in turn affected the alongshore transport rates by 73%, representing 385% of the annual littoral drift of the Dutch Coast. For the same conditions, RUE simulations were reduced to 72% in order to match the beach envelopes and consequently had a lower impact on the alongshore sediment transport (56%).

Thus, by improving the parameterization of near-bed wave orbital motion there is less need to (over-)calibrate sediment transport. The skewness-asymmetry parameterization also proved to be robust, in the sense that predictions were less sensitive to variations in the user-defined calibrations factors. Therefore, RUE parameterization results in a closer coupling of cross-shore and alongshore sediment transport in the nearshore, which improves the long-term hindcast and expectations related to complex coastal and estuarine environments.

Supplementary Materials: Supplementary materials can be found at http://www.mdpi.com/2077-1312/7/6/188/s1.

Author Contributions: Conceptualization, M.B.A., G.R. and M.G.K.; Formal analysis, M.B.A.; Funding acquisition, M.G.K.; Investigation, M.B.A.; Methodology, M.B.A., G.R. and H.R.A.J.; Resources, M.G.K.; Software, H.R.A.J.; Supervision, M.G.K.; Writing—original draft, M.B.A.; Writing—review and editing, G.R., H.R.A.J. and M.G.K.

Funding: This research was funded by ERC Consolidator agreement 647570 to MGK.

Acknowledgments: Code development was supported by Deltares and we gratefully thank Dirk-Jan Walstra for the arrangements. We acknowledge Rijkswaterstaat and USACE-FRF for making field data available. Authors appreciated the fruitful discussions and model assistance from Bart Grasmeijer and Pieter Koen Tonnon. Delft3D source code is freely distributed and available at the Deltares (SVN) repository. The authors encourage the compilation and usage of this version. The supplementary materials contains the SVN path along with specific instructions to activate the options between IH and RUE parameterization. Please contact the corresponding author for further technical information and consult Delft3D-Deltares web-page for compiling and general model instructions.

Conflicts of Interest: The authors declare no conflict of interest.

Appendix A. Parameterizations of Orbital Velocities

Appendix A.1. Isobe Horikawa (IH)

The equations of IH presented here were based on the Delft3D source code and van Rijn [42]. As a first step, the velocity amplitude U_w is estimated based on linear wave theory applying local wave conditions: wave height (H_{rms}), period (T) and local water depth (h)

$$U_w = \frac{H_{rms}\,\pi}{T\,\sinh(kh)} \tag{A1}$$

where, k is the wave number calculated from the dispersion relation: $w^2 = gk\,\tanh(kh)$; g is the gravity acceleration and w the angular frequency. Then, the maximum velocity amplitude U_{max} is computed according to

$$U_{max} = 2U_w\left[-0.4\left(\frac{H_{rms}}{h}\right) + 1\right] \tag{A2}$$

From the velocity amplitude, the maximum onshore (U_{on}) and offshore (U_{off}) directed velocities are calculated following

$$u_{on} = U_{max}\left(0.5 + (r_{max} - 0.5)\,\tanh\left(\frac{r_a - 0.5}{r_{max} - 0.5}\right)\right)$$
$$u_{off} = U_{max} - u_{on} \tag{A3}$$

where,

$$r_a = \begin{cases} -5.25 - 6.1\,\tanh\left(A_1\,\dfrac{U_{max}}{\sqrt{(g/h)}} - 1.76\right) & \text{if } \quad r_a \geq 0.5 \\[2mm] 0.5 & \text{if } \quad r_a < 0.5 \end{cases} \tag{A4}$$

$$A_1 = -0.0049\,(T\sqrt{g/h})^2 - 0.069\,(T\sqrt{g/h}) + 0.2911$$

and,

$$r_{max} = 0.62 < -2.5\,(h/L) + 0.85 < 0.75 \tag{A5}$$

Then, the duration of onshore (T_{for}) and offshore (T_{back}) directed velocities are calculated in

$$T_{for} = \frac{u_{off}}{u_{on} + u_{off}}\,T$$
$$T_{back} = T - T_{for} \tag{A6}$$

And finally the onshore $u_{on}(t)$ and offshore $u_{off}(t)$ directed velocities are computed with t varying from 0 to wave period T according to

$$u(t) = \begin{cases} u_{on} \sin\left(\pi \dfrac{t}{T_{for}}\right) & \text{for} \quad t < T_{for} \\ -u_{off} \sin\left[\dfrac{\pi}{T_{back}}(t - T_{for})\right] & \text{for} \quad t \geq T_{for} \end{cases} \tag{A7}$$

Appendix A.2. Ruessink (RUE)

The Ruessink method description is based on Ruessink et al. [18]. The method starts with the calculation of Ursell number (U_r) as

$$Ur = \frac{3\sqrt{2}\,H_{rms}}{8}\frac{k}{(kh)^3} \tag{A8}$$

Based on the Ursell number the total non-linearity (B) and the phase (Ψ) are computed as the following

$$B = p_1 + \frac{p_2 - p_1}{1 + \exp\dfrac{p_3 - \log(Ur)}{p_4}} \tag{A9}$$

$$\Psi = -90° + 90° \tanh\left(\frac{p_5}{Ur^{p_6}}\right) \tag{A10}$$

where: $p_1 = 0$; $p_2 = 0.857$; $p_3 = -0.471$; $p_4 = 0.297$; $p_5 = 0.815$; $p_6 = 0.672$;

With the total non-linearity and phase, the skewness (Sk) and asymmetry (As) are calculated with

$$Sk = B\cos(\Psi) \tag{A11}$$

$$As = B\sin(\Psi) \tag{A12}$$

Then a new non-linearity (r) and phase (ϕ) are derived from B and Ψ as in

$$b = \frac{\sqrt{2B^2}}{9 + 2B^2}$$
$$r = \frac{2b}{1 + b^2} \tag{A13}$$

$$\phi = -\tan^{-1}\left(\frac{A}{S}\right) - \frac{\pi}{2} = -\Psi - \frac{\pi}{2} \tag{A14}$$

The amplitude of orbital velocities U_w is obtained following Equation (A1). Finally, $u(t)$ is calculated based on the velocity amplitude, total non-linearity and phase, as demonstrated in

$$u(t') = U_w f \frac{\sin(\omega t') + \dfrac{r\sin(\phi)}{(1+f)}}{1 - r\cos(\omega t' + \phi)} \tag{A15}$$

where, $f = \sqrt{(1 - r^2)}$ is a dimensionless factor to match the amplitude of u and U_w. In addition, at Equation (A15), t is modified into t' to ensure $u(0) = 0$.

$$t' = t - \left[\frac{1}{\omega}\arcsin\left(\frac{r\sin(\phi)}{1+f}\right)\right] \tag{A16}$$

References

1. Elgar, S.; Guza, R.T. Nonlinear model predictions of bispectra of shoaling surface gravity waves. *J. Fluid Mech.* **1986**, *167*, 1–18. [CrossRef]
2. Zijlema, M.; Stelling, G.; Smit, P. SWASH: An operational public domain code for simulating wave fields and rapidly varied flows in coastal waters. *Coast. Eng.* **2011**, *58*, 992–1012. [CrossRef]

3. Malej, M.; Mith, J.M.; Salgado-Dominguez, G. *Introduction to Phase-Resolving Wave Modeling with FUNWAVE*; ERDC/CHL CHETN-I-87; US Army Corps of Engineers: Washington, DC, USA, 2015.

4. Van der Spek, A.J.; Beets, D.J. Mid-Holocene evolution of a tidal basin in the western Netherlands: A model for future changes in the northern Netherlands under conditions of accelerated sea-level rise? *Sediment. Geol.* **1992**, *80*, 185–197. [CrossRef]

5. Van Rijn, L. *Sand Budget and Coastline Changes of the Central Coast of Holland between Den Helder and Hoek van Holland, Period 1964–2040*; Deltares: Delft, The Netherlands, 1995.

6. Beets, D.J.; van der Spek, A.J.F. The Holocene evolution of the barrier and the back-barrier basins of Belgium and the Netherlands as a function of late Weichselian morphology, relative sea-level rise and sediment supply. *Neth. J. Geosci.* **2000**, *79*, 3–16. [CrossRef]

7. Stive, M. A model for cross-shore sediment transport. In Proceedings of the 20th International Conference on Coastal Engineering, Taipei, Taiwan, 9–14 November 1986; pp. 1550–1564.

8. Dean, R.; Perlin, M. Intercomparison of near-bottom kinematics by several wave theories and field and laboratory data. *Coast. Eng.* **1986**, *9*, 399–437. [CrossRef]

9. Rienecker, M.M.; Fenton, J.D. A Fourier approximation method for steady water waves. *J. Fluid Mech.* **1981**, *104*, 119–137. [CrossRef]

10. Isobe, M.; Horikawa, K. Study on Water Particle Velocities of Shoaling and Breaking Waves. *Coast. Eng. Jpn.* **1982**, *25*, 109–123. [CrossRef]

11. Roelvink, J.; Brøker, I. Cross-shore profile models. *Coast. Eng.* **1993**, *21*, 163–191. [CrossRef]

12. Van Rijn, L.; Walstra, D.; Grasmeijer, B.; Sutherland, J.; Pan, S.; Sierra, J. The predictability of cross-shore bed evolution of sandy beaches at the time scale of storms and seasons using process-based Profile models. *Coast. Eng.* **2003**, *47*, 295–327. [CrossRef]

13. Walstra, D.; Reniers, A.; Ranasinghe, R.; Roelvink, J.; Ruessink, B. On bar growth and decay during interannual net offshore migration. *Coast. Eng.* **2012**, *60*, 190–200. [CrossRef]

14. Roelvink, J.; Stive, M. Bar-generating cross-shore flow mechnisms on a beach. *J. Geophys. Res. Oceans* **1989**, *94*, 4785–4800. [CrossRef]

15. Ruessink, B.G.; Kuriyama, Y.; Reniers, A.J.H.M.; Roelvink, J.A.; Walstra, D.J.R. Modeling cross-shore sandbar behavior on the timescale of weeks. *J. Geophys. Res. Earth Surf.* **2007**, *112*. [CrossRef]

16. Grasmeijer, B. Process-Based Cross-Shore Modeling of Barred Beaches. Ph.D. Thesis, Utrecht University, Utrecht, The Netherlands, 2002.

17. Dubarbier, B.; Castelle, B.; Marieu, V.; Ruessink, G. Process-based modeling of cross-shore sandbar behavior. *Coast. Eng.* **2015**, *95*, 35–50. [CrossRef]

18. Ruessink, B.; Ramaekers, G.; van Rijn, L. On the parameterization of the free-stream non-linear wave orbital motion in nearshore morphodynamic models. *Coast. Eng.* **2012**, *65*, 56–63. [CrossRef]

19. Warner, J.C.; Sherwood, C.R.; Signell, R.P.; Harris, C.K.; Arango, H.G. Development of a three-dimensional, regional, coupled wave, current, and sediment-transport model. *Comput. Geosci.* **2008**, *34*, 1284–1306. [CrossRef]

20. Villaret, C.; Hervouet, J.M.; Kopmann, R.; Merkel, U.; Davies, A.G. Morphodynamic modeling using the Telemac finite-element system. *Comput. Geosci.* **2013**, *53*, 105–113. [CrossRef]

21. Bertin, X.; Oliveira, A.; Fortunato, A.B. Simulating morphodynamics with unstructured grids: Description and validation of a modeling system for coastal applications. *Ocean Modell.* **2009**, *28*, 75–87. [CrossRef]

22. Nardin, W.; Fagherazzi, S. The effect of wind waves on the development of river mouth bars. *Geophys. Res. Lett.* **2012**, *39*. [CrossRef]

23. Nienhuis, J.H.; Ashton, A.D. Mechanics and rates of tidal inlet migration: Modeling and application to natural examples. *J. Geophys. Res. Earth Surf.* **2016**, *121*, 2118–2139. [CrossRef]

24. Luijendijk, A.P.; Ranasinghe, R.; de Schipper, M.A.; Huisman, B.A.; Swinkels, C.M.; Walstra, D.J.; Stive, M.J. The initial morphological response of the Sand Engine: A process-based modelling study. *Coast. Eng.* **2017**, *119*, 1–14. [CrossRef]

25. Nardin, W.; Fagherazzi, S. The Role of Waves, Shelf Slope, and Sediment Characteristics on the Development of Erosional Chenier Plains. *Geophys. Res. Lett.* **2018**, *45*, 8435–8444. [CrossRef]

26. Tonnon, P.; Huisman, B.; Stam, G.; van Rijn, L. Numerical modelling of erosion rates, life span and maintenance volumes of mega nourishments. *Coast. Eng.* **2018**, *131*, 51–69. [CrossRef]

27. Grunnet, N.M.; Walstra, D.J.R.; Ruessink, B. Process-based modelling of a shoreface nourishment. *Coast. Eng.* **2004**, *51*, 581–607. [CrossRef]

28. Briere, C.; Giardino, A.; van der Werf, J. Morphological modeling of bar dynamics with DELFT3D: the quest for optimal free parameter settings using an automatic calibration technique. *Coast. Eng. Proc.* **2011**, *1*, 60. [CrossRef]

29. Storms, J.E.A.; Stive, M.J.F.; Roelvink, D.J.A.; Walstra, D.J. Initial Morphologic and Stratigraphic Delta Evolution Related to Buoyant River Plumes. In Proceedings of the Coastal Sediments '07, New Orleans, LA, USA, 13–17 May 2007. [CrossRef]

30. Edmonds, D.A.; Slingerland, R.L. Significant effect of sediment cohesion on delta morphology. *Nat. Geosci.* **2010**, *3*, 105–109. [CrossRef]

31. Guo, L.; van der Wegen, M.; Roelvink, D.J.; Wang, Z.B.; He, Q. Long-term, process-based morphodynamic modeling of a fluvio-deltaic system, part I: The role of river discharge. *Cont. Shelf Res.* **2015**, *109*, 95–111. [CrossRef]

32. Van der Vegt, H.; Storms, J.; Walstra, D.; Howes, N. Can bed load transport drive varying depositional behaviour in river delta environments? *Sediment. Geol.* **2016**, *345*, 19–32. [CrossRef]

33. Braat, L.; van Kessel, T.; Leuven, J.R.F.W.; Kleinhans, M.G. Effects of mud supply on large-scale estuary morphology and development over centuries to millennia. *Earth Surf. Dyn.* **2017**, *5*, 617–652. [CrossRef]

34. Geleynse, N.; Storms, J.E.; Walstra, D.J.R.; Jagers, H.A.; Wang, Z.B.; Stive, M.J. Controls on river delta formation; insights from numerical modelling. *Earth Planet. Sci. Lett.* **2011**, *302*, 217–226. [CrossRef]

35. Nahon, A.; Bertin, X.; Fortunato, A.B.; Oliveira, A. Process-based 2DH morphodynamic modeling of tidal inlets: A comparison with empirical classifications and theories. *Mar. Geol.* **2012**, *291–294*, 1–11. [CrossRef]

36. Olabarrieta, M.; Geyer, W.R.; Kumar, N. The role of morphology and wave–current interaction at tidal inlets: An idealized modeling analysis. *J. Geophys. Res. Oceans* **2014**, *119*, 8818–8837. [CrossRef]

37. Nienhuis, J.H.; Ashton, A.D.; Nardin, W.; Fagherazzi, S.; Giosan, L. Alongshore sediment bypassing as a control on river mouth morphodynamics. *J. Geophys. Res. Earth Surf.* **2016**, *121*, 664–683. [CrossRef]

38. Deltares. Delft3D-FLOW: Simulation of multi-dimensional hydrodynamic flows and transport phenomena, including sediments. In *User Manual*; Deltares: Delft, The Netherlands, 2017.

39. Booij, N.; Ris, R.C.; Holthuijsen, L.H. A third-generation wave model for coastal regions: 1. Model description and validation. *J. Geophys. Res. Oceans* **1999**, *104*, 7649–7666. [CrossRef]

40. Ris, R.C.; Holthuijsen, L.H.; Booij, N. A third-generation wave model for coastal regions: 2. Verification. *J. Geophys. Res. Oceans* **1999**, *104*, 7667–7681. [CrossRef]

41. Deltares. SVN Repository. 2018. Available online: https://svn.oss.deltares.nl/repos/delft3d/tags/delft3d4/7545 (accessed on 18 June 2019).

42. Van Rijn, L.C. *Principles of Fluid Flow and Surface Waves in Rivers, Estuaries, Seas, and Oceans*, 2011st ed.; Acqua: Amsterdam, The Netherlands, 1990; ISBN 978-90-79755-02-8.

43. Van Rijn, L.C.; Walstra, D.J.R.; van Ormondt, M. *Description of TRANSPOR2004 and Implementation in Delft3D-ONLINE*; Final Report; Deltares (WL): Delft, The Netherlands, 2004.

44. Van Rijn, L.C. Unified View of Sediment Transport by Currents and Waves. I: Initiation of Motion, Bed Roughness, and Bed-Load Transport. *J. Hydraul. Eng.* **2007**, *133*, 649–667. [CrossRef]

45. Van Rijn, L.C. Unified View of Sediment Transport by Currents and Waves. II: Suspended Transport. *J. Hydraul. Eng.* **2007**, *133*, 668–689. [CrossRef]

46. Van Rijn, L.C.; Walstra, D.J.R.; van Ormondt, M. Unified View of Sediment Transport by Currents and Waves. IV: Application of Morphodynamic Model. *J. Hydraul. Eng.* **2007**, *133*, 776–793. [CrossRef]

47. Abreu, T.; Silva, P.A.; Sancho, F.; Temperville, A. Analytical approximate wave form for asymmetric waves. *Coast. Eng.* **2010**, *57*, 656–667. [CrossRef]

48. Soulsby, R.; Hamm, L.; Klopman, G.; Myrhaug, D.; Simons, R.; Thomas, G. Wave–current interaction within and outside the bottom boundary layer. *Coast. Eng.* **1993**, *21*, 41–69. [CrossRef]

49. Walstra, D.J.R.; van Rijn, L.C.; van Ormondt, M.; Briere, C.; Talmon, A.M. The Effects of Bed Slope and Wave Skewness on Sediment Transport and Morphology. In Proceedings of the Coastal Sediments '07, New Orleans, LA, USA, 13–17 May 2017. [CrossRef]

50. Rijkswaterstaat. The Yearly Coastal Measurements (in Dutch: De JAarlijkse KUStmetingen or JARKUS). 2017. Available online: https://opendap.deltares.nl/thredds/catalog/opendap/rijkswaterstaat/jarkus/caralog.html (accessed on 18 June 2019).

51. Wijnberg, K.M.; Terwindt, J.H. Extracting decadal morphological behaviour from high-resolution, long-term bathymetric surveys along the Holland coast using eigenfunction analysis. *Mar. Geol.* **1995**, *126*, 301–330. [CrossRef]

52. Donald K.; Stauble, M.A.C. Sediment dynamics and profile interactions: DUCK94. In Proceedings of the 25th International Conference on Coastal Engineering, Orlando, FL, USA, 2–6 September 1996; pp. 3921–3934.

53. Trowbridge, J.; Young, D. Sand transport by unbroken water waves under sheet flow conditions. *J. Geophys. Res. Oceans* **1989**, *94*, 10971–10991. [CrossRef]

54. Gallagher, E.L.; Elgar, S.; Guza, R.T. Observations of sand bar evolution on a natural beach. *J. Geophys. Res. Oceans* **1998**, *103*, 3203–3215. [CrossRef]

55. Walstra, D.; Hoekstra, R.; Tonnon, P.; Ruessink, B. Input reduction for long-term morphodynamic simulations in wave-dominated coastal settings. *Coast. Eng.* **2013**, *77*, 57–70. [CrossRef]

56. Benedet, L.; Dobrochinski, J.; Walstra, D.; Klein, A.; Ranasinghe, R. A morphological modeling study to compare different methods of wave climate schematization and evaluate strategies to reduce erosion losses from a beach nourishment project. *Coast. Eng.* **2016**, *112*, 69–86. [CrossRef]

57. Bailard, J.A. Modeling on-offshore sediment transport in the surf zone. In Proceedings of the 18th International Conference on Coastal Engineering, Cape Town, South Africa, 14–19 November 1982; pp. 1419–1438.

58. Baar, A.W.; de Smit, J.; Uijttewaal, W.S.J.; Kleinhans, M.G. Sediment Transport of Fine Sand to Fine Gravel on Transverse Bed Slopes in Rotating Annular Flume Experiments. *Water Resourc. Res.* **2018**, *54*, 19–45. [CrossRef]

59. Baar, A.; Albernaz, M.B.; van Dijk, W.; Kleinhans, M. The influence of transverse slope effects on large scale morphology in morphodynamic models. *E3S Web Conf.* **2018**, *40*, 04021. [CrossRef]

60. Dubarbier, B.; Castelle, B.; Ruessink, G.; Marieu, V. Mechanisms controlling the complete accretionary beach state sequence. *Geophys. Res. Lett.* **2017**, *44*, 5645–5654. [CrossRef]

61. Bailard, J.A. An energetics total load sediment transport model for a plane sloping beach. *J. Geophys. Res.* **1981**, *86*, 938–1095. [CrossRef]

62. Van Thiel de Vries, J. Dune Erosion During Storm Surges. Ph.D. Thesis, Delft University of Technology, Delft, The Netherlands, 2009.

63. Brinkkemper, J.A.; Aagaard, T.; de Bakker, A.T.M.; Ruessink, B.G. Shortwave Sand Transport in the Shallow Surf Zone. *J. Geophys. Res. Earth Surf.* **2018**, *123*, 1145–1159. [CrossRef]

MDPI
St. Alban-Anlage 66
4052 Basel
Switzerland
Tel. +41 61 683 77 34
Fax +41 61 302 89 18
www.mdpi.com

Journal of Marine Science and Engineering Editorial Office
E-mail: jmse@mdpi.com
www.mdpi.com/journal/jmse

9 783039 284849